Construction Management

Rajiv Abhyankar

VISHWAKARMA PUBLICATIONS
VP

Construction Management

First Edition – October 2015
© Rajiv Abhyankar

ISBN 978-93-83572-91-5

Published by:
Vishwakarma Publications
283, Budhwar Peth, Near City Post,
Pune 411 002.
Phone No: 020 24448989 / 20261157
Email: info@vpindia.co.in
Website: www.vpindia.co.in

Cover Design, Typeset and Layout
Chaitali Nachnekar - Vishwakarma Publications

Printed at
Repro India Limited, Mumbai.

FOREWORD

Over the years, business philosophy has changed and at the same time work environment has also changed. Gone are the days when Industries spent huge amount on training new recruits. Recruits spent anywhere between 6 months to 2 years in Training Centres learning various aspects of Business and work culture. Employees looked at the employment as life long career till retirement and leaving an organisation never crossed the minds of majority of employees. Only a few would leave an organisation to start a business on their own or go abroad for higher education or lucrative salary. Now the organisations look at training as wasteful expenditure and not as an investment. At the same time employees look at the organisation as a vehicle that they board to get down at some destination as a step towards their career goal. Working lifetime in an organisation does not fit in their ethos. Every new entrant in an organisation is expected to be knowledgeable and aware of the expectations from him. Industry wants Universities to produce graduates who are ready to work like ready to eat food. This concept overlooks the basic objective of University education that is to give general awareness of engineering disciplines and make students explore on their own specific details and if necessary follow higher education in a particular specialisation. The creation of new engineering colleges with or without adequate resources and need to show better results to attract more students has its effect on quality of engineers passing out from these institutions.

This has created a situation where the new entrant is confused and the lower & middle management is frustrated with the staff.

This book is a humble effort to give persons starting their careers in construction an introduction to the task ahead.

TABLE OF CONTENTS

CONSTRUCTION MANAGEMENT

The words **Construction Management** evoke a feeling that it is something related to Civil Engineering. This is far from true. In modern business all field activities that are part of Installation process are clubbed under **Construction Management**. It is different from other management streams and the managers here have to be more self-reliant as they are away from specialized support in day to day situations. In a factory, there is well organized set up to help the line managers to deal with Industrial Relation, Public Relations, Legal and other issues on the spot. In most organisations, the Construction Manager or Site Manager is representing the Chairman of the company and he has to take a broader view of situations before taking any action or making any commitment.

In a factory environment, the workforce is recruited by HR following company policy and basic education qualifications are strictly adhered to for every appointment. In the field, most of workforce has hardly gone to school and do not possess systematic training from any institute barring few categories like welders. Skill sets are developed by hands on experience. The Construction Manager has to depend on his judgment to employ such persons by his previous experience with the persons or on input from reliable sources. There is always pressure from various quarters to accommodate persons who do not meet basic criteria or not interested in doing any kind of work. This includes project affected people, persons from nearby villages and persons known to influential local persons. This affects the per capita work output.

This is a discipline where a person needs to remember his science lessons from school and basic engineering principles. At every step, someone will raise doubts and the practicing construction professional needs to give technically and logically correct answer. He needs to be in touch with latest management principles and apply them. He should understand financial matters and work towards a positive cash flow in difficult situations.

India being federal country, there are regulations enacted by both Central and respective State Governments. The regulatory norms and requirements differ from state to state. Construction companies need to keep themselves updated with these requirements and take suitable actions. In this aspect also Construction Management is different from other disciplines of management.

Construction Management is the overall planning, co-ordination and control of a project construction from inception to completion aimed at meeting a client's requirements in order to produce a functionally and financially viable project that will be completed on time within authorized cost and to the required quality standards & legal requirements.

1.1   Definition:

Construct means to form by assembling or combining parts; build.

Construction means "The act or process of constructing".

1.2   Construction Process:

It encompasses Mechanical erection, Electrical work, Controls and Instrumentation related works and last but not least Civil works

1.3   Construction Management Processes:

- **Marketing**
- **Planning and Budgeting**
- **Project Management**
- **Commercial & Contract Management**

- **Design & Drafting**

- **Procurement**

- **Quality Assurance & Control**

- **Execution**

- **Document Control**

- **Health & Safety**

- **Stores/ Material Management**

- **Maintenance**

- **HR**

- **Administration and Public Relations**

- **Legal**

- **Security**

- **Finance & Accounts**

The processes are explained below

1.4 Marketing

Any organisation requires marketing to quote against tenders and get orders. Marketing group has to periodically meet existing and potential customers to find out their feedback on current project progress and future order potential. It is also required to keep abreast of developments in the market and keep management informed for taking suitable decisions.

1.5 Construction Planning

The Planning process is carried out at 2 levels

> Corporate or Head Office level – Macro planning

> Site level – micro planning

Planning is an important function that requires intricate knowledge of various processes involved, resources required for the process and performance standards for the activities involved in the processes. With this knowledge only realistic schedules can be developed for a project.

1.6   Construction Planning (Macro)

> **■   Based on Long Term Plan and Business forecast following are planned**

- ➤ Manpower Induction, Training
- ➤ Equipment procurement or hiring
- ➤ Technology acquisition
- ➤ Funding requirement

Every organisation has a vision, mission and goals. The organisation strives to achieve these in a structured manner. Organisations prepare Strategic plans and Long Term Plans to achieve the vision. These plans are reviewed by top management periodically and actions are taken.

An organisation has to work in an environment that may not be friendly. It has to take stock of changes in government policies and their effect on the organisation. This includes changes in tax rates and changes in legislations. It also needs to keep itself abreast with changes in economic scenario and its effect. With economic liberalization, the impact of developments in other countries, also affect organisations.

New developments are taking place in every field and new products get introduced in market every year. Customers are also looking at getting latest technology and product designs in products they purchase. Every organisation needs to be ready with latest techniques and assimilate the same. Organisations need to have their own research and development, but a construction company depends on products designed and manufactured by a different organisation. Construction companies do not have their own research and development department.

Construction companies need to have agreements with manufactures to know new developments and be a part of the design team during drawing release so that easy constructability can be ensured. New products may require new processes and these need to be acquired and mastered. Similarly by participating in industrial exhibitions & trade fairs, new developments in tools and equipments can be understood for implementing within the organisation. This helps in improving productivity.

Equipment procuring needs to be planned in advance as some of these are long lead items and require more than 1 year to get delivered. Many types of equipment need to be imported and necessary actions have to be taken. Financial analysis for Purchase vis a vis Hire needs to be done.

Collaboration agreements, is one way of acquiring new technology, but this has financial implications. An organisation may not have personnel with requisite skills to use the new technology and processes. For this, training is one way of upgrading skills. Other way is to recruit persons with requisite skills.

All the above need adequate cash flow and the organisation needs to work out its funding requirement both long term and short term. Negotiations with Financial institutes need to be carried out for getting loans on reasonable terms.

- **Based on Orders on hand and schedule requirements, the following is planned from one Project to another**

  ➢ Transfer of Manpower

  ➢ Transfer of Equipment

  ➢ Transfer of Material

- **Based on Orders on hand, schedule requirements and availability, the following is planned**

  ➢ Induction of Manpower

  ➢ Induction of Equipment

  ➢ Induction of Material

  ➢ Site Infrastructure

The general practice is to prepare histograms for the project with respect to resource requirements and prepare action plans. By superimposing the histograms of different projects, resources can be identified for transfer from one project to another project. If gaps are identified the same are filled by new recruitment and purchase of material & equipments.

Site Infrastructure needs to be planned based on the project requirements. There are standards for office space requirement per person and using these, office accommodation needs to be arranged. With introduction of portable furnished containers for office, storage, kitchen and toilets, many companies now use these instead of building temporary structures that need to be cleared on project completion. Infrastructure needs to be compliant with regulatory requirements.

■ Generate Plans and MIRs

At macro planning level, the plans are

- ➤ Yearly plans for 2 or 3 years
- ➤ Quarterly plan for the current year
- ➤ Month-wise plans

In Plan Do Check Act (PDCA) cycle, the output / performance needs to be checked with plan so that Management can take corrective action. This is achieved by generating Management Information Reports (MIRs). MIRs are generated based on the requirements of Management. MIRs cover every aspect of operations including manpower, material, equipment, safety, quality, billing, cash-flow and productivity.

1.7 Construction Planning (Micro)

■ Based on schedule requirement and availability, the following is planned

- ➤ Induction/ Transfer of Manpower
- ➤ Induction / Transfer of Equipment
- ➤ Induction / Transfer of Material

Induction of manpower can be done either from Head Office or authorised locally. Local induction is mostly for unskilled jobs and sometimes for specific skills like computer operators. This meets the aspirations of local population from the project.

■ Study Construction Drawings and work out strategy for Construction

This is a very important aspect of the planning function at site. Construction drawings need to be studied for clarity and any inadvertent mistakes. This avoids delays in execution. Based on the drawing, required material, equipment and skilled manpower is identified and arranged to suit the schedule. Action to expedite inputs can be initiated based on the study.

Strategy for Construction includes designating approaches and exits for safe movement of men and material. It also includes selection of Construction equipment, and its layout for safe movement of material and workforce.

- **Generate Project level plans, schedules & MIRs**

Project level plans and schedules include

> ➢ Monthly Plan and Schedules
>
> ➢ Weekly Plan and Schedules
>
> ➢ Daily Plan and Schedules
>
> ➢ Shift wise Plan and Schedules

Based on these plans and schedules corresponding MIRs are generated.

1.8   Construction Project Management

- **For a Construction company, it is essential to organize a separate Project Management group to handle the process for their scope of work.**

A Construction company need not worry about allotting weightages and Work Breakdown Structure (WBS) for the project modules. It has to ensure that sufficient time is given to perform its scope of work and the inputs are available at appropriate time in the schedule.

For a Project schedules are prepared at different levels designated as L1, L2, and L3 and so on with different levels of details. L1 is higher level schedule and its requirements need to be addressed by subsequent levels of schedules. Construction Company gets involved in the schedules of L2 level and below levels.

Primavera project planner and Microsoft Project are two popular commercially available softwares, which are used in the industry. Suretrack has facilities to integrate data with Primavera and is used for lesser number of activities. For daily schedule Excel sheets are also used. Choice of software depends on preference of client.

1.9 Commercial & Contract Management

After marketing has secured an order and signed contract, the contract becomes Commercial & Contract Management's responsibility to execute and complete the contract.

Every Contract has Terms of Payment that indicate how the contractor would be paid during the course of contract. Terms of Payment are divides as

- Initial Advance – The advance paid on fulfilling certain conditions as laid down in Tender/ Contract. These include

  - Signing of Contract

  - Submission of Bank Guarantee/ Corporate Guarantee

  - Submission of Contract Performance Guarantee

  - Submission of Schedule

- Mobilization Advance – The conditions are laid down in Contract to define meaning of Mobilization. These include

  - Deployment of specified Construction equipment at site

  - Deployment of specified manpower to site

  - Development of specified minimum Site infrastructure

- Advance against receipt of material – This is typically a feature of Civil construction. Advance is paid against receipt of cement, aggregate, steel and other material specified in contract

- Progressive Payment – This payment is released against the work done at site. The payment terms include

  - o Civil

    - Bar bending

    - Form work

    - Reinforcement binding

    - Pouring of Concrete

- - Excavation
  - Back filling
  - Brickwork
  - Mechanical
    - Placement
    - Alignment
    - Welding
    - Radiography
    - Hydraulic Test
    - Non Destructive Testing
  - Electrical
    - Placement and alignment
    - Cable tray erection
    - Junction box erection
    - Cable laying
    - Termination
    - Testing
    - Trial run of equipments
  - Controls and Instrumentation
    - Testing and calibration of instruments
    - Placement
    - Impulse tubing
    - Cable tray erection
    - Instrument rack erection
    - Junction box erection
    - Cabling

- Termination
- Testing

➢ Milestone linked Payment – This payment is released against achieving specified Milestones as per contract. These include

- Hydraulic Test
- Light up of Boiler
- Oil flushing completion of Turbo generator
- Steam blowing completion
- Turbine on Barring gear
- Rolling of Turbine
- Synchronization
- Coal Firing
- Safety Valve floating
- Trial run completion
- Performance Guarantee test completion

➢ Final Payment – This payment signifies the closing of contract. It is released after closing all outstanding issues. This payment is released after obtaining equivalent amount Bank Guarantee for duration of warrantee period.

- **Price break up finalization**

This is important in any contract as it defines how cash inflow would take place during the course of the contract. Every client wants to delay cash outflow whereas every contractor wants to get higher cash inflow during initial stages. It is important to strike a balance so that work at site is not affected.

If at tender stage item wise price break up is captured, then there is not much scope in price break up finalisation. However where the details are not submitted at tender stage, this detailing is crucial. Contractor is expected to submit his proposal and then after discussions the price break up schedule gets finalised and approved. Billing can be done against approved price breakup schedule or pro rata as it is called in some contracts.

▪ **Invoicing and Cash collection**

In Construction Industry the bill for work done is also called Running Account (RA) Bill. This is because every bill cumulates the quantities so far completed and is against progress in a billing cycle.

The steps involved are

> ➤ Taking measurements – Quantities in Cubic metres, Square metres, metres, Kilograms, numbers, sets and Metric tonnes etc.

> ➤ Review and approval of measurements

> ➤ Entering into Measurement Book (MB)

> ➤ Getting measurements certified and approved

> ➤ Generating a bill based on approved measurements

> ➤ At Invoice stage, applicable taxes are added on 100% of the current billing cycle work. From the Gross value of Bill, proportionate deductions are made for progressive payment already received, advances and Milestone linked payments to arrive Nett Invoice value

This Invoice along with certified measurements is submitted to the client. The Invoice also needs to be accompanied by proofs of complying with statutory requirements as stated in Contract. Technical group of client forwards the Invoice along with MB to Finance group of client after scrutiny and approval. Finance group of client after verification sends the Invoice to Cash section within Finance for releasing the payment.

Every Contract specifies cycle time for releasing payment after receipt of Invoice. What one should understand is that this cycle time from the date the Invoice is accepted after scrutiny. It is therefore important to follow up at every stage of Invoice processing and make it payable at the earliest. Invoice processing has effect on Working Capital of a company and hence much importance is given to this process.

Cash collection is the process of getting Invoices processed at client office, giving clarifications for any queries during verification, keeping track of any additional deductions made by client and getting the cheques in hand or ensuring funds transfer by electronic means. It also involves collecting Tax Deducted at Source (TDS) certificates from client and handing these to Finance group in company.

Many Contracts have a provision to issue fresh Bank Guarantee of reduced value during the course of contract, as the advance gets recovered in each RA Bill. This helps in reducing the overall outstanding Bank Guarantee amount of a company and it can request for fresh Bank Guarantees without enhancing its limit with the banks. This also helps in reducing the Bank charges being paid by a company.

▪ **Conflict resolution**

When two or more parties work together, there are bound to conflicts that need to be resolved amicably. This mainly arises from difference in perception before contract signing and understanding of persons executing the contract. Every contract provides for a mechanism to address this issue and elaborates on the actions to be taken by affected parties.

The conflicts relate to

➢ Technical – Related to specifications

➢ Contractual – Related to

- Scope of work
- Terminal Points definitions
- Schedule
- Progress of work
- Quality
- Safety
- Meeting Statutory requirements
- Exclusions from Contract
- Release of payments
- Persons posted at site

Contract Managers have to deal with each conflict and resolve it amicably without affecting commercial interests of the company. They may take the help of Marketing Executives dealing with the order to get insight into the issue and in some cases taking them for a meeting with client to clarify on specific issues where clarity is required.

Contract Managers have a difficult job. They have to protect Client's interest in internal meeting and protect Company interest during interaction with clients. They need to have thorough knowledge of specifications and the contract.

1.10   Design & Drafting

- **Prepare specifications and drawings for Site Infrastructure**

Every construction project has different requirements regarding

- ➢ Construction Power
- ➢ Construction water
- ➢ Drinking water
- ➢ Site office and cabins
- ➢ Residential accommodation for staff and labour
- ➢ Temporary roads and drains
- ➢ Storage space
- ➢ Facilities

Considering the above, specifications need to be drawn and drawings/ schemes made for procurement, fabrication and construction of the infrastructure.

- **Prepare design and drawings for Temporary supports, structures and fixtures**

The manufacturer supplies only material that is billed to the customer and material required for facilitating erection works is not supplied. Based on the product to be erected/ constructed designs need to be developed for preassembly beds, scaffolds, working platforms and temporary supports etc. In certain cases, the design calculations need to be submitted to consultant/ customer for their approval when it involves critical activities.

- **Prepare As Built Drawings**

During construction, it is observed that work cannot be performed as per approved drawing/ document. This may involve reinforcement sizing and spacing, depth & size of excavation, piping routing, cable routing, termination details of terminal blocks, signal destination and so on. Necessary approvals are taken to proceed with the work and construction/ commissioning completed. For future reference and usage all these deviations need to be marked on drawings, cable schedules, logic diagrams and circuit diagrams. This is a major activity post construction/ commissioning and as per contract these documents are required for closing the contract.

- **Prepare excavation drawings if needed**

Excavation drawings are prepared and issued as a normal practice by Engineering department. However in multinational companies this responsibility lies with the site. Depending on the site conditions, construction group prepares excavation drawings.

1.11 Procurement

This process is part of Material management function and is described in detail in the chapter on Material Management.

- **Purchase of Product and Services as per schedule**
- **Coordinate for inclusion of vendors in customer's approved vendor list**
- **Coordinate for submission and approval of Vendor drawings, documents and schedules**
- **Coordinate for Inspection at Vendor works**

1.12 Quality Assurance & Control

Quality Assurance is focused on providing confidence that quality requirements will be fulfilled. It is the systematic measurement, comparison with standard, monitoring of processes and associated feedback loop that confers error prevention.

Quality Control is focused on fulfilling quality requirements. It is focused on Process output.

- **Prepare, issue and review of Quality Management System documentation**

- **Every organisation has its own Quality Management system and strives to get it certified to ISO standards for business purpose. To fulfil this, the related documentation is prepared, issued and updated.**

- **Preparation of Quality Plans**

- **Quality Plan is a document specifying which procedures and associated resources shall be applied by whom and when to a specific project, product, process or contract.**

There are two types of Quality Plans in any organisation

- ✓ Standard Quality Plan

  - This document is requirement of Quality Management System as per ISO 9000 standards

  - This is used in assuring a prospective customer that a robust Quality Assurance system is in practice

  - It details the inspection stages, criteria and the authority for inspection/ acceptance

This document takes into account

  - ➢ Engineering requirement of product

  - ➢ Applicable National / International standards and their requirement.

  - ➢ Applicable Statutory requirement

  - ➢ Industry practices

  - ➢ Available resources

- ✓ Contract specific Quality Plan

  - This is a modified version of Standard Quality Plan, incorporating Contract specific Quality requirements as agreed.

  - This details the stages where Customer representative will participate and his role.

▪ Field Quality Assurance Program

This is a Quality plan for site activities. It also provides log sheets to be filled to demonstrate conformance to requirements.

Field Quality log sheets have provision to record design requirement, tolerance limits, values observed at manufacturing works and finally the values observed at site during construction. Filled up log sheets are handed over to customer for future reference.

For rotating equipment it is essential to obtain the shop documentation before starting alignment at site.

▪ Preparation Design mix for concrete

At every project a design mix for concrete is prepared with local ingredients and submitted for approval to customer. The details of design mix are explained in civil work activities.

▪ Inspection at vendor works

Inspection is carried out at vendor works as per Purchase order conditions and approved Quality plan. Customer representative is associated with the inspection if necessary.

▪ Inspection of work at site

This is an important activity and sufficient workforce needs to be deployed. It involves stage inspection and preparation of quality records associated with the activities.

▪ Interface with Customer/ Vendor on quality related issues

QA has to resolve quality related issues with Customer/ Vendors. It has to coordinate with Engineering for getting approvals on deviations and communicate with Customer/ vendors.

▪ Quality Month celebration

November month is observed as Quality month. QA&C organise various competitions, activities and lectures related to quality in this month.

> ▪ **Conduct Quality Audits and Reports**

Periodic audit is essential to maintain health of Quality Management system. QA has the responsibility to conduct periodic audits and monitor resolution/ corrective actions. It has to coordinate with external auditors. Based on audits and quality related issues, reports are generated and submitted to management.

1.13   Execution

> ▪ **Prepare, maintain and issue Manuals, Standard Operating Procedures and Work Instructions**

Every product needs an erection manual describing the steps and precautions to be taken during erection. These manuals are prepared after discussing with the product designers the salient points and specific requirements. Any change in product design or technology would necessitate review of associated manual and corrections as required.

Standard operating procedures and work instructions need to be issued for maintaining safety and quality of work. Based on incidents and near miss reported by sites, a review of these is required for their continued suitability and modifications required.

> ▪ **Prepare activity wise manpower and productivity standards. Monitor the actual progress vis a vis the resources deployed**

Margins are always under pressure and it is necessary to keep variable costs down. To achieve this, standards need to be developed and implemented. It is important to monitor deployment of resources and corresponding progress of work. If any mismatch is there, corrective action needs to be taken.

> ▪ **Execute work as per schedule, quality standard and Health & Safety requirements**

Doing work as per contract schedule is of utmost importance but maintaining quality standards with proper Health and Safety precautions cannot be compromised.

- **Prepare maintain and issue Job Safety Analysis**

Job safety analysis needs to be carried out for all activities and operation controls decided. This needs to be reviewed and revised based on feedback and incidence/ accident reports from sites.

- **Establish and maintain Site infrastructure**

A proper site infrastructure is essential for meeting project schedule. Depending on the feedback from marketing, some advance actions need to be taken before receipt of order. Then only it is feasible to meet the requirements of project. In addition to establishing the infrastructure, it needs to be maintained in good condition. Contracts for maintaining the facilities also need to be awarded and monitored.

- **Meet Statutory requirements at Project**

There is need to identify project specific statutory requirements and take action to meet them. Often a directory of concerned offices with addresses, names of persons and their contact details is prepared for every project.

- **Customer interface at Project site**

The construction team at site faces the customer on daily basis. The team is expected to keep in touch with HO and keep track of drawing release/ revisions, manufacturing and dispatch status. Since the team works on customer premises, it is hard to hide anything from customer. Many times deficiencies are noticed by customer before others and less response time is available for contacting HO or Manufacturing / Engineering for probable solutions. Every issue needs to be tackled tactfully without affecting company interest.

- **Interface with local bodies and population at Project site**

Smooth work at project site depends on the cooperation of population of surrounding villages and the local bodies. Village roads are not designed for movement of heavy vehicle traffic. Roads can get damaged due to this heavy load movement and there is possibility of accidents as these roads are narrow. Many villages put restriction on heavy vehicle movement and fix timings for the same to avoid accidents.

Many of the villagers are affected by project having sold their land and expect employment opportunities during construction phase. They also expect petty contracts for various supplies / services required at project site.

Renovation of school buildings, provision of furniture, lights and fans in classrooms, toilets for school and supply of computers are some of the things the local bodies expect from Project developer as well as EPC contractor.

These issues need to be reviewed and action taken for smooth work at site.

- **Identify Construction/ Testing equipment requirement as per schedule and arrange**

Based on the products to be erected, customer facilities, site location and project schedule, the construction/ testing equipments are identified. Deployment schedule is prepared and the equipment is arranged by way of transfer/ new purchase/ hire or put in contractor's scope.

- **Ensure fitness of equipment before use**

This is a legal as well as safety requirement. List of competent persons and authorised agencies is maintained state wise for use at project sites. Fitness test is conducted at site before first time use and also as required by law. Records are maintained for inspection by authorities.

- **Ensure calibration of equipment before use**

All testing instruments need to have valid calibration certificate by NABL approved laboratory before use for proper use. National Accreditation Board for Testing and Calibration Laboratories (*NABL*) is an autonomous body under the aegis of Department of Science & Technology of Government of India. Even brand new equipment needs to be calibrated before use if the manufacturer's laboratory is not certified. Directory of NABL approved testing laboratories with addresses, contact person details and capabilities need to be maintained for use by project sites. Some instruments can be calibrated at specific places only and the agency has long waiting list. In view of this, action needs to be taken sufficiently in advance so that work is not affected.

HO should issue schedule of calibration to all project sites and monitor compliance.

> ▪ **Estimate Power & water requirement**

Power and water requirement for a project is not constant over the period of contract. It varies as per schedule of activities and resource deployment. Projection of Month wise/ Week wise daily requirement of both power and water need to be worked out and submitted to customer. Customer has to arrange these accordingly.

For power requirement calculation, the number of equipments in use per shift and their hours of operation with ratings are taken as basis. For water daily concreting quantity with curing requirement are taken as basis.

> ▪ **Tool Box meeting**

Tool Box meeting is an effective tool in communicating concerns and building teamwork. It needs to be conducted before start of work at the beginning of shift. The target for the team during the shift, safety hazards involved in the work and precautions to be taken are explained by the engineer/ supervisor. It is essential to keep record of the tool box meeting and obtain signatures of all workers that they have understood the requirements of the job including safety precautions. In case of incident/ accident, tool box meeting records are taken as evidence that management has performed its obligation.

1.14 Document Control

> ▪ **Prepare, receive, store, issue and retrieve documents as per QMS guidelines**
> ▪ **Ensure availability of pertinent documents at work spot**

1.15 Health & Safety

There has to be a separate department dealing with Health & Safety both at HO and project site reporting directly to Head of organisation. Many organisations have their Health and Safety system and get it certified to OHSAS 18001 standard.

> ▪ **Prepare, maintain and issue Health and Safety manual**

The Health and Safety manual describes the system in place and how the requirements of OHSAS 18001 standard are met. It contains the Occupational Health & Safety policy of the organisation. The manual is revised based on feedback, audit findings and revision to International standard. It is a practice now to publish the manual on internal/ external web site of the organisation for use of all stake holders.

- **Prepare, maintain and issue Project specific Health and Safety manual**

This is a contractual requirement of most of the contracts. Project specific manual addresses local issues and legal requirements. A sample manual is annexed for ready reference.

- **Prepare, maintain and issue Project specific Emergency Preparedness Plan**

This is a contractual requirement of most of the contracts. Project specific Emergency preparedness plan takes into account local conditions and details role, responsibility and action to be taken by the site staff including customer personnel. A sample plan is annexed for ready reference.

- **Provide assistance in Job Safety Analysis preparation**

Execution group has to identify the activities/ processes used in the organisation. For each of these activities/ processes Job safety analysis needs to be carried out with possible variations. Based on these, operational controls are defined and issued as Operational Control Procedures (OCP) or Safe work practices. Safety group provides key inputs in preparation of Job Safety analysis.

- **Induction training on Health & Safety at every location of the organisation for every employee, worker and visitor**

The induction training is important to familiarise all stakeholders about the policy of organisation and their responsibility in following safety requirements. Different modules of training are prepared keeping in mind the audience and job requirement. It is customary to have separate modules for executives, supervisors, customer employees, contractor executives/ supervisors, workers and visitors. Separate modules are also prepared based on area of work like civil, mechanical etc.

Duration of training varies as per site condition, but statutory authorities expect the training to be of more than 2 days for workers.

▪ Ensure implementation of Safe work practices

Safe work practices need to be published/ displayed at appropriate locations. These have to be in languages understood by majority of workforce. These are explained to workers during tool box meetings as part of implementation. Safety engineers/ stewards have to monitor activities with respect to safe work practices and bring to the notice of concerned in case of deviation.

▪ Interface with Statutory authority for Health & Safety

This is an important activity as the representatives of statutory authority visit periodically the project site for inspection. Health & Safety department at site would interact with them and brief them on the status. Requisite reports are prepared and submitted by the site safety team. Site safety team would accompany the officials during site visit and inform the site team in advance of the visit.

▪ Issue and maintain work permit system

Work permit system is explained in detail in a separate chapter. During the construction phase, it is the responsibility of safety team at site to issue and control work permits.

▪ Inspection of Construction equipment, tools & tackles and vehicles

These need to be inspected periodically to ensure safe working. Any defects noticed will be brought to the notice of the concerned and the equipment would be removed from work till it is repaired/ replaced.

▪ Ensure Health checkup of employees/ workers as per Statutory requirement

Pre-employment health checkup is mandated by law. Subsequently periodic health checkup of workers is also required to ensure that working at site has not adversely affected the worker's health. Persons permitted to work at height need to be certified for doing so. The health records need to be shown to government officials during their site visit.

- **Install and maintain Fire extinguishers**

Fire extinguishers need to be installed at various locations in site. These have to be selected on type of fire and size of the area to be covered. Indian Standards give guidelines for selection of fire extinguishers and the numbers required. In addition to this buckets for pouring water and some buckets filled with sand also need to be strategically placed.

Inspection of fire extinguishers and periodic replacement is taken care by safety team. Safety team arranges for training in fire safety and also arranges mock fire drill along with security team.

- **Tool Box meeting**

Safety engineer/ steward should enable holding of Tool Box meeting daily. They should preserve the records.

- **Issue and control Personnel Protection Equipment (PPE)**

The safety team at site is responsible to issue requisite PPE to staff and workers. Basic PPE i.e. Helmet and Safety shoes are issued just after induction training when gate pass is issued to an individual. Visitor pass is also issued after safety training and issuing PPE to visitor. Further PPE like gloves, gum boots etc are issued to workers as per job requirement. Safety team gives indent for purchase of PPE.

- **Conduct Mock drills**

Mock drills are essential to check preparedness for meeting emergencies. Safety team conducts mock drills with the help of security and suppliers in use of fire fighting and for rescue operation.

- **Site Safety Committee and Safety meetings**

Involvement of all stakeholders is important to create safe working environment. For this a site safety committee is formed with representation of all contractors and their labour. Specific issues are brought out for discussion and views obtained. Site safety committee also undertakes joint safety inspection and find out deviations from safe work practice. The meeting is conducted once a month.

> ■ **Safety week celebration**

First week of March every year is observed as Safety week. Various programs related to safety like quiz, posters competition, slogans competition etc are held. Workers are encouraged to participate in these competitions and also to speak in front of gathering. Awards and certificates are presented to best contractor, best worker etc based on safety performance during the year. Some sites recognise workers on monthly basis to inculcate safety awareness.

> ■ **Audit and generate MIRs**

Any system is effective if its implementation is periodically audited for compliance. Safety team has a schedule for periodic inspection and audit of construction activities. Periodic reports are generated for management to take decision on the health of safety system.

1.16   Stores/ Material Management

The process is explained in detail in Material Management chapter.

> ■ **Receipt, storage, preservation and issue of material as per QMS**
>
> ■ **Organize material receipt inspection and raise nonconformance report as applicable**
>
> ■ **Take up with supplier/ Insurance company for defective material**
>
> ■ **Follow up with transporters for delivery/ transport of material**
>
> ■ **Procurement/ Issue of consumables**
>
> ■ **Material reconciliation with customer**

1.17   Maintenance

This is an important function and directly affects the profitability of an organisation. Maintaining all equipment in good running condition is very important. It has to maintain register of all equipment with their current location, spares for these equipment and expenditure incurred on individual equipment since induction.

▪ Prepare and follow maintenance schedule for equipment

Every equipment manufacturer gives periodic checks and maintenance schedule for the equipment supplied. Based on these a schedule is prepared and communicated to user departments. Ensure checks and maintenance is carried out as per schedule.

▪ Arrange for repair and maintenance of equipment

Depending on the scope of work and number of equipments deployed, at many sites a separate repair facility is created to carry out minor repairs of the equipment. However the repairs that require special facilities are outsourced and contracts awarded for the same. Similarly overhaul of equipment is carried out by agencies authorised by Original Equipment Manufacturer (OEM).

▪ Hire equipment and augment capability in emergencies or for critical activity

Inspite of taking all precautions, equipment can breakdown affecting project schedule. The equipments are deployed taking into account normal workload and sometimes there are spurt in activities and the available equipments cannot meet the short time higher demand. To meet such exigencies, equipment is taken on hire for short duration. This can happen when there is sudden heavy rainfall and water draining becomes a constraint. This can also happen when mass concreting is done for major foundations requiring additional transit mixers for a day or two. If too many trailers arrive on a single day the unloading capacity of existing crane becomes inadequate and additional cranes need to be hired.

▪ Spare parts inventory management

Spare parts for equipment need to be maintained for 15 to 20 years. These need to be moved from one project site to another along with the equipment. On consumption, of spare parts the same need to be ordered and received.

▪ Expenditure history for all equipment

History card needs to be maintained, for each equipment giving details of its deployment, running hours, spares consumed and expenditure on repairs. This facilitates taking a decision on retaining or disposing equipment.

> ■ **Disposal of outdated and beyond economic repair equipment**

After prolonged use, spare parts for some equipment are not available in market and the cost of repairs is more than the depreciated value of the equipment. Under these conditions the equipment is labelled as Outdated or Beyond Economic Repair (BER) and action taken to dispose it off.

> ■ **Generate MIRs**

Periodic Management Information Reports are generated giving details of

> ➤ Adherence to maintenance schedule

> ➤ Repairs carried out and cost of repairs

> ➤ Consumption of spare parts

> ➤ Progress of work for outsourced activities

> ➤ Equipment identified for disposal and disposed

1.18   HR

Human Resources or Personnel as the department was known earlier, has the following functions.

> ■ **Implementation of Company HR policy**

This includes

> ➤ Maintenance of Service records of employees

> ➤ Maintenance of Leave records

> ➤ Compiling of Appraisal reports and conducting promotion process.

> ➤ Issue of Promotion/ Transfer/ Termination orders

> ➤ Resolving Industrial relations issues

> ➤ Group insurance, Provident Fund, Gratuity and other related issues.

▪ **Induction and training of manpower**

Based on the skill gap projections, HR takes action for recruitment and conducts interviews for selection of suitable individuals. Based on the recommendation of selection committee appointment letters are sent. On joining the individuals are given induction training making them aware of company policies and details of the organisation. Training curriculum varies according to the grade/ trade the individual is selected for.

Apart from induction training, individuals are deputed for acquiring higher skills based on recommendation of management. Training needs are identified by individuals as well as their controlling officers.

▪ **HR related Statutory compliance**

Returns need to be filed periodically to government authorities. HR department compiles the information and files the returns.

▪ **Generate MIRs**

HR activities related periodic reports are generated and submitted to management.

1.19   Administration and Public Relations

In most organisations, Administration is attached to HR. Public relations is a separate department at HO, but at site Administration is given the responsibility to maintain good public relations.

▪ **Arrange accommodation and transport for staff**

The staff posted at project site needs residential accommodation. Guest house needs to be maintained for visiting officials and bachelor hostel for individuals staying alone. Hotel accommodation also needs to be arranged for visitors. All these persons have to be transported to site and back. Administration takes care of this work. The lease agreements and rent payment are the responsibility of administration team. Verification of vehicle logs and processing the bills is taken care by administration team.

▪ Arrange catering services at site, guest houses and bachelor hostels

---

Administration arranges catering services at these locations to ensure hygienic food is served. If catering service is not available, cooks and helpers are engaged to prepare food.

▪ Arrange office stationery and consumables

Administration has the responsibility to maintain sufficient stock of stationery and issue it to users.

▪ Upkeep of Office, Guest Houses and staff colony

The facilities need to be kept clean and administration has to engage labour to keep facilities clean and hygienic. Administration has to arrange supporting staff to carry out work.

▪ Upkeep and maintenance of Labour Colony

Sanitation in and water supply to Labour Colony are critical issues. (Site work can stop anytime if these are not maintained properly.) Administration has to take steps to ensure proper living conditions in Labour colony. It also has to take care of allotment of accommodation to various contractors. With the help of security peace and order has to be maintained in the colony. Unauthorised shops need to be prevented from operating in the labour colony.

▪ Issue and control employee identity cards, gate passes and maintain attendance registers

---

Every employee needs to be issued identity card and also gate pass for entering the site. Contractors and their workers are issued separate gate passes with photographs. When an employee leaves the organisation, his identity card and gate pass needs to be collected back before settling his dues. Same process is followed for contractor workers who leave site. Control is exercised at gate for counting labour entering and leaving site for collecting daily attendance figures.

> ▪ **Tie up with local Hospital for medical tests and emergency treatment**

This is important activity as the number of persons undergoing tests is high. With a tie up, employees can get medical assistance without hassles. In case of accidents, emergency treatment is given without making advance payments. Bills need to be settled expeditiously so that the tie up continues.

> ▪ **Comply with Labour related legal requirements and file returns**

Administration has to ensure that all contractors are complying with statutory requirements. Administration has to ensure remittances as per statutory requirements. Administration has to maintain registers and records as specified by laws and file periodic returns to government authorities.

> ▪ **Interact with local population**

Goodwill of local population is important for project progress. Most of the projects have only one road connecting the project to highway. Any disturbance can stop total movement for the project. Administration does the following

> ➤ Participate in community development projects around the Project

Digging bore wells for drinking water, renovating school buildings, construction of toilets, providing furniture and fixtures to schools etc are some of the activities that are undertaken.

> ➤ Maintain good relations with people in surrounding villages

Interacting with population and addressing their concerns is one way of better relations with local population. Providing them suitable employment opportunity is also required.

> ▪ **In case of incidence/ accident take necessary action**

Taking injured person immediately to local hospital for treatment is essential. In case of reportable accident, administration has to inform authorities in prescribed formats. Necessary actions to avoid industrial relation problem is also a responsibility of administration.

- **Maintain Primary Health centre**

Every project site requires a primary health centre. It depends on the provisions in contract. Many customers put the responsibility on EPC contractor. In such case primary health centre is set up. It should be manned by paramedics and should be available round the clock. It should have a refrigerator to store emergency medicines and antidotes for snakebite. An ambulance also should be available round the clock. Arrangement for a doctor to visit also should be made for consultation if full time doctor is not available.

- **Generate MIRs**

Necessary periodic reports are generated and submitted to management.

1.20   Legal

Every organisation requires a legal department to advise on legal matters. Contractual clauses need to be studied for their implications before signing the contract. The Legal cell has following responsibilities

- **Registration of Project site with Statutory authorities**

Project sites need to be registered for

> - Provident Fund
> - Excise duty
> - Sales Tax
> - Value Added Tax
> - Labour licence
> - Professional Tax
> - Work contract Tax

- **Ensure legal compliance**

Ensure Legal compliance for the above.

- **Filing of periodic returns**

Every legal requirement requires filing of periodic returns. These are filed by legal cell.

- **Assist in litigations involving company**

Every organisation has to face litigations. The Legal has to identify suitable advocate for representing the company. During the case proceedings Legal has to provide necessary supporting documents to the advocate and represent the company in courts.

- **Generate MIRs**

Requisite reports are generated for information of management.

1.21   Security

The contract specifies the role of EPC contractor. In some contracts, customer operates Gate security and internal security is the responsibility of EPC contractor. In Greenfield projects, the entire security responsibility is given to EPC contractor with skeleton security arrangement for owner's office by customer.

The first step is to study the plant layout and understand vulnerabilities. Based on this threat analysis is carried out and points identified to eliminate/ reduce threats. A project specific security plan is developed and submitted to customer for approval. A sample plan is annexed.

There are agencies providing security services and these are registered with government and have a license to provide security services. Deployment of guards depends on the area to be covered and stage of project implementation.

- **Implement Project Security plan**

Chief Security officer and his team have the responsibility to implement the project security plan.

- **Protection of company property**

Theft from the project site and damage to material needs to be prevented. At gate every person and vehicle needs to be checked to ensure no company material is taken out clandestinely. Similarly persons and vehicles need to be checked while coming in for entry of restricted items.

- **Interface with local police**

Security group has to keep good relations with local police, so that adequate police force is available at short notice when trouble erupts. Lodging of complaints for thefts and handing over anti-social elements to police is also the responsibility of security group. Since it is only a watch and ward service, care is required to be taken while apprehending culprits. Security guards cannot manhandle any person.

- **Generate MIRs**

Security plan details out the reports expected from security group.

1.22 Finance & Accounts

These are basically two different functions. Finance deals with cashflow management while Accounts deals with book keeping. Finance has to ensure adequate funds for operations and deal with Banks for obtaining credit facilities.

- **Implement company finance policy**

Every organisation has its stated finance policy and delegation of authority. Finance and Accounts have to ensure that the policy is implemented and there is compliance with statutory regulations.

- **Arrange and monitor Bank Guarantees**

A company gives and receives many Bank guarantees or corporate guarantees. Finance has to interact with Banks and get adequate limits for getting guarantees issued. The guarantees already issued need to be collected back and handed over to issuing bank. In case the validity of a bank guarantee needs to be extended, the same has to be arranged. The bank guarantees received from vendors/ contractors need to be monitored for their validity and action taken to keep them valid till the contracts are closed. Bank guarantees need to be returned or encashed, on advice of concerned departments.

- **Participate in Award of Work/ Purchase order process**

Finance and Accounts are members of Purchase committee. They have to ensure that the company policy is followed and vendors/ contractors are financially sound before awarding them work.

- **Maintain accounts**

This involves passing bills and releasing payment. The books of accounts are maintained as per company policy.

One important function is to pay salaries to employees on time and remit taxes to government. Provident fund and other deductions from salary also need to be remitted within specified time limit.

Taxes need to be deducted from vendor's bill and remitted to treasury account. Tax deducted at Source (TDS) certificates need to be issued to vendors/ contractors.

- **Assist in financial audits**

There are three types of financial audits

- ✓ Internal audit
- ✓ Statutory audit (External auditors)
- ✓ Government audit

Internal audit is carried out by company's audit team to ensure compliance with company policy.

As per law, external auditors have to conduct audit of company's books and certify the accounts. This is called statutory audit.

Government audit is conducted by Comptroller and Accountant General (CAG). Though it is for public sector companies, there is a demand that CAG should audit companies providing services on Public Private Partnership (PPP) model or serving infrastructure services.

▪ Generate MIRs

Periodic Financial reports are generated and submitted to management. Financial results are compiled periodically and published as per legal requirement.

▪ Document Control

Document has the same meaning as defined in ISO-9001. Document Control covers following activities for the documents:

A) Preparation, Review, and Approval

Responsibility and authority for these activities for documents used, needs to be defined at company level. (At site only as built documents are prepared.)

B) Receipts and Storage

On receipt of documents, the details are entered in Document Control Register (DCR). The location where the document is stored is entered in DCR for easy retrieval. Racks and Cabinets in adequate quantity are required in Documentation room. A table and chairs also need to be provided for persons who desire to scan through the documents without getting them issued.

Documents received in electronic media also need to be controlled in same way.

C) Issue – Documents are issued to Engineers, supervisors and contractors as per distribution list issued by owner of the documents.

D) Retrieval- Document control system should be such that, it should be possible to retrieve any document in shortest possible time.

E) Revision (includes addenda) - Revision control system varies in every company. Before start of Project, revision indexing needs to be finalised between the parties involved and applied to project specific documents. There has to be a system in vogue to inform project site of impending revision to a document, so that work does not proceed beyond certain specified point.

F) Removal of obsolete documents- Use of obsolete document plays havoc in projects. On receipt of a revised document, it is essential to remove all old documents from work spots and issue revised documents.

G) Maintenance of revisions- Older revision may be kept for reference by originator. Older revisions need to be removed from all other users.

H) Disposal – Obsolete and torn documents need to be disposed off. Normal practice is to burn these documents in a pit under supervision to avoid their falling in wrong hands.

1.23   Execution

To achieve the desired progress, adequate site infrastructure needs to be arranged in time. Site infrastructure consists of

- **Site offices (Permanent & Portable cabins)**

There are standard office space requirements per person. In addition to this, conference halls, library, document room, pantry, dining space and communication room space needs to be planned. Engineers and supervisors need to be near work spots and for them portable cabins need to be provided. These portable cabins are available in market and do not need elaborate foundations. These are fully furnished and require a power and water connection to become fully operational.

- **Security office and cabins**

A Security office near gate to accommodate staff and documents is required. In addition to this at strategic points security cabins are placed to keep watch. Watch towers are also erected at strategic points. The size and number depends on the site size and terrain.

- **Civil Laboratory**

A civil laboratory with testing equipments is required when civil works are in the scope of work. A curing tank of adequate size is required to ensure proper curing of test cubes.

- **Radiography source pit and room**

This has to be constructed as per Atomic Energy Regulatory Board (AERB) guidelines. The drawing needs to be submitted to AERB for approval. Location needs to be carefully decided.

- **Store sheds (Closed, Semi closed & Air conditioned)**

The space requirement depends on the scope of work, number of units, type of material, storage instructions by suppliers and place of origin for supplies. This also depends on the space allotted by customer for storage. At any given time 3 months stock would be at site for smooth progress. Care has to be taken during layout preparation that there are no safety hazards and proper approach is provided for movement of material.

- **Store yards**

Store yards space is decided as per storage instruction by suppliers and availability of area given by customer. The store yards need to be graded and provided with all weather internal roads and drains to ensure that there is no water stagnation. All weather approach road is required from gate to storage yards. The yard should have fencing and security gate. Wooden or concrete sleepers for stacking material are required. Store yards are divided by grid marking; A to Z in one direction and 1 to 99 in perpendicular direction to alphabetical division. This way storage location of any material can be recorded in stock register for easy retrieval.

- **Fabrication Yard**

If scope of work includes fabrication, then fabrication yards need to be planned. Fabrication yards need to be graded and provided with all weather internal roads with drains. Fabrication yards should be adjacent to raw material storage with material handling facility. Power outlets and earthing grid need to be provided in fabrication yard. Fabrication yards should be fenced with security gate.

For civil works, fabrication involves bending of reinforcement bars or rebars as these are called. In addition to this inserts need to be fabricated.

Mechanical fabrication attracts levy of excise duty and Fabrication yard needs to be registered with Excise authorities.

- **Batching Plants**

Size and number of batching plants depend on scope of work, output requirement and redundancy requirement. If sufficient space is not available, sourcing of ready mix concrete needs to be explored.

| ▪ | **DG Sets** |

DG sets are required for ensuring uninterrupted power supply for continuity of work. DG sets are also required if proper voltage is not available from construction power source. The number and location of DG sets has to be decided as per plot plan and consumption points. Commissioning of DG sets requires permission from Electrical inspector and excise duty needs to be paid on power generated.

| ▪ | **Construction Power scheme** |

Construction needs power. Greenfield projects are in rural area where the supply is erratic and unreliable. The power requirement during construction is more than 2 MW and for this a robust network needs to be created. It consists of transformers, DG sets, earth pits, switchgear rooms, distribution boards and cabling. Construction power needs to be created and submitted to both customer and electrical inspector for approval.

| ▪ | **Construction water scheme including storage facility** |

Construction needs huge quantity of water. Customer is responsible to provide water at one point, but he need to be informed of the daily consumption. Water needs to be stored and then distributed to work spots. This can be done by either pumping water by laying pipelines to individual area sumps or can be transported by tankers.

| ▪ | **Repair/ Maintenance work shop** |

It is required to take up minor repairs to equipments.

| ▪ | **IT facilities (PCs, Laptops, scanners, printers and plotters)** |

Construction drawings are sent electronically to remote site, thereby reducing the time cycle. Necessary IT infrastructure needs to be established from day one.

Local Area Network within the site needs to be established for connecting different offices.

- **Internet connection**

Drawing transmittal requires higher bandwidth. Suitable internet connection needs to be established. Temporary arrangements with local agency can be made for downloading drawings till internet connection is arranged at site.

- **Primary Health Centre**

Primary Health centre is required at site to provide first aid. It should have a refrigerator to store medicines, blood pressure checking equipment, stretcher and room for diagnosis. It is manned by paramedics.

- **Labour Colony**

Depending on the scope of work and estimated manpower deployment, labour colony is constructed with water supply, power and sanitation facilities.

- **Drinking water scheme**

Safe drinking water supply is mandated by law and arrangements need to be made to provide drinking water to work spots, offices and labour colony.

- **Temporary roads and drains**

Drawings for permanent roads and drains get released after finalising the layout and water discharge volume requirements. For initial work, temporary roads and drains need to be constructed.

- **Temporary illumination and area lighting**

Permanent illumination system gets finalised after a long time. During construction phase, adequate lighting is required for working round the clock. For this temporary illumination and area lighting needs to be designed and established. The power source can be construction power and for emergencies and remote areas by mobile diesel generator set operated lighting system.

- **Temporary Fire protection equipment**

Adequate measures need to be taken to prevent fire. Fire extinguishers of different types and small tankers need to be kept ready for containing any fire.

- **On site Facilities for Labour as per statutory requirements**

These include toilets and urinals, canteen, shelters and crèche for children.

1.24   Descriptive Questions

1. Describe the various processes involved in Construction Management

2. Describe the steps in a International Competitive bid finalisation upto order placement

3. What is Job Safety Analysis? How does it help in construction

4. List the activities of Safety department at site

5. Describe the process of Equipment identification and deployment for a project

6. What is Job Safety Analysis? How does it help in construction

7. Why Design and drafting section is required in a Construction company? Describe activities of the section.

8. Why does a Construction company need a Project Management department?

9. What is a Tool Box meeting? Why it should be held before start of work? Describe the purpose and content of the Tool Box meeting.

10. Describe various ways to increase productivity in an organisation.

1.25   Multiple Choice Questions

State True or False

SL	Statement	True	False
01	Public relations is important for timely execution of a project		
02	Interface with villages surrounding the project is not required		
03	There is requirement to employ new staff for new project		
04	Histograms for all projects are required		
05	Project Management is not required for a construction company		
06	Inventory levels have no impact on Productivity of an organisation		
07	Quality Plans detail stages of inspection		

LIFE CYCLE OF A PROJECT

2.1   What is a Project?

A Project is a temporary endeavor with defined beginning and end undertaken to meet unique Goals and Objectives. It is often time constrained and constrained by Funding or Deliverables or Resources.

This means it is not a routine activity.

Every project has defined stages and most of the projects follow a set path. Be it a quality improvement project or setting up a new manufacturing facility following steps will be involved. Since we are explaining the concept with a (thermal power) project in mind, certain steps would not be necessary in a project that is not related to power plant.

▪ Concept

Every project starts with an idea or a concept. The individual conceiving the idea needs to work on that and make it a saleable idea/ concept. Marketing the idea/ concept is very important.

▪ Internal Approvals

Any project needs resources to implement it. Resources are not freely available and there is a cost associated with resources. In any organisation, approvals are required to launch a project. For obtaining the approval, scope of project, team members, time frame and resources need to be defined. Cost benefit analysis is also required to get approval. Any organisation would invest in a project only if Return on Investment is lucrative.

▪ Launching of Project

Once the management of organisation approves the project, it is officially launched. The team members, scope, boundary, time schedule and cost of project are defined and announced.

▪ Appointment of Consultant

Appointment of external consultant may not be necessary for all internal projects, however for specific projects where internal expertise is not available, consultant is appointed. For large infrastructure project, appointment of consultant is necessary. Consultant prepares the specification of the project keeping in view latest technological trends and applicable regulations. It is not necessary to appoint a single consultant for the project. There can be number of consultants involved in a single project with every consultant helping in some specialized area while the main consultant concentrates on the project specifications. It is a practice to appoint separate consultant for obtaining environmental clearance, other consultant for transmission line related works, third consultant for land related issues, separate consultant for railway related matters and so on.

▪ Detailed Project Report

Consultant would prepare a detailed project report outlining the scope of project, sources of inputs like water, coal etc and detailed estimate for various components of the project with time frame. The detailed project report is required to be submitted for obtaining external approvals.

▪ External Approvals

For setting up a project, approvals from local state government and central government are required.

▪ Power Purchase Agreement (PPA)

For a power project, power purchase agreement is essential. It ensures that the power generated would be sold at reasonable rate. PPA has to be as per Central Electricity Regulatory Commission (CERC) or State Electricity Regulatory Commission guidelines.

▪ Fuel Supply Agreement (FSA)

For running life of the power project, assured fuel supply at reasonable rate is required. Many gas based power projects are idle because gas is not available at affordable rate. FSA defines the mine from which coal would be supplied to the project. Mega power projects can develop their own coal mines after bidding for the same.

■ **Water Supply Tie up**

A thermal power project needs huge quantity of water. It is essential to have a long term arrangement for water supply. For some projects, developing captive catchment area becomes essential. There are instances where state governments have assured water supply to many projects from the same source without ascertaining the availability. One needs to exercise caution in this regard, as in case of scarcity first priority would be given to drinking water and agriculture.

■ **Land acquisition**

Central Electricity Authority (CEA) has issued guidelines regarding requirement of land for a thermal project based on capacity of the proposed project. These need to be followed.

Earlier state governments used to help in acquiring land for the projects. However private developers are required to acquire land at market rate.

■ **Environmental Clearance**

This is very important for any infrastructure project. Land acquisition for the project is a prerequisite for seeking Environmental clearance. Documentary proof of intent by land owners to sell their land for project is required.

There are conditions for granting Environmental clearance by Central government that are categorized as

    (i)   Conditions for pre-construction phase

    (ii)   Conditions for construction phase

    (iii)   Conditions for post-construction/ operation phase

    (iv)   Conditions for entire life of the project

Environmental Impact Assessment (EIA) is an important management tool for ensuring optimal use of natural resources for sustainable development. EIA has now been made mandatory under the Environmental Protection Act, 1986 for 29 categories of developmental activities involving investments of Rs. 50 crores and above.

The process is illustrated below

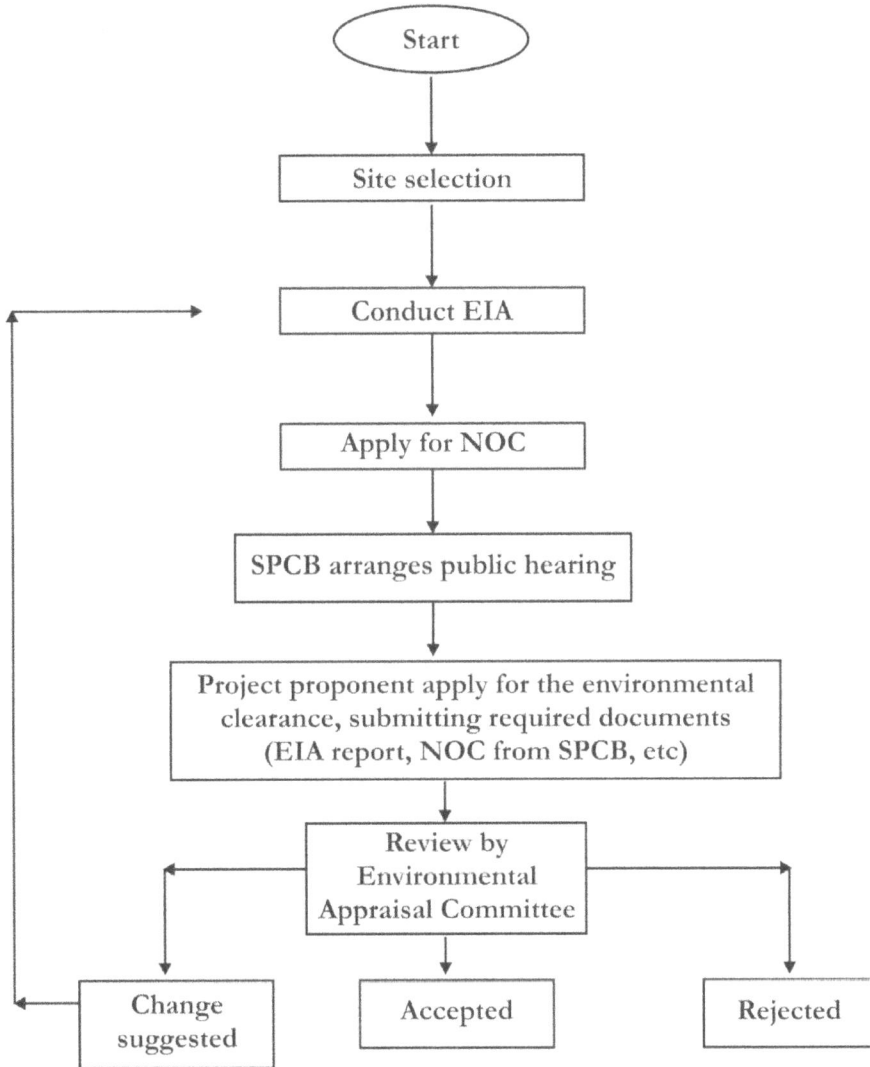

```
                        ( Start )
                            |
                            v
                   +-----------------+
                   |  Site selection |
                   +-----------------+
                            |
                            v
                   +-----------------+
      +----------->|   Conduct EIA   |
      |            +-----------------+
      |                     |
      |                     v
      |            +-----------------+
      |            |  Apply for NOC  |
      |            +-----------------+
      |                     |
      |                     v
      |      +----------------------------+
      |      | SPCB arranges public hearing|
      |      +----------------------------+
      |                     |
      |                     v
      |   +-------------------------------------------+
      |   | Project proponent apply for the environmental |
      |   | clearance, submitting required documents      |
      |   | (EIA report, NOC from SPCB, etc)              |
      |   +-------------------------------------------+
      |                     |
      |                     v
      |   +--------------------------------+
      |   |          Review by             |
      |   |       Environmental            |
      |   |    Appraisal Committee         |
      |   +--------------------------------+
      |        |            |            |
      |        v            v            v
      |  +----------+  +----------+  +----------+
      +--| Change   |  | Accepted |  | Rejected |
         | suggested|  +----------+  +----------+
         +----------+
```

Note:

EIA – Environmental Impact Assessment

NOC – No Objection Certificate

TOR – Terms of Reference for EIA

SPCB – State Pollution Control Board

- **Vendor Finalization**

Based on the specifications drawn by consultant, bids are invited for various packages and work awarded.

- **Power Evacuation Finalization**

The power generated from projects needs to be evacuated and transmitted to distant consumers. Power Grid operates the national grid while individual states have their own transmission companies.

Private power producers need to connect their power to either state grid or national grid based on the power generated. For larger plants, power is connected to national grid. For this agreement needs to be made and location of substation where the power would get connected is finalized. A bay in the substation would be reserved for individual power producer. Wheeling charges are required to be paid for transmitting power.

Transmission line is to be erected from power plant to the identified substation after carrying out survey and obtaining route clearance. The same line would provide start up power for initial commissioning.

- **Schedule Finalization**

Schedules with different levels of details are prepared for the project. Consultant gives Level 1 (L1) schedule with specifications and bid document. After vendor finalisation Level 2 (L2) and Level 3 (L3) schedules are submitted by vendors for their scope of work. After discussions and necessary modifications these schedules are approved for base line monitoring. Approved schedules are submitted to financial institutions also for their monitoring of the project.

- **Financial Closure**

The stage when the project developer ties up with the banks/financial institutions for the funds required for the project, and conditions precedent to initial drawing of the debt have been fulfilled.

This is an important stage and most of the bidders to a project are anxious to know the status. Once financial closure is achieved, vendors are assured of timely payment against progress of work.

- **Basic Engineering**

Basic engineering defines the sizing of various equipments, which is required for giving guaranteed output. It is submitted by main contractor as Design Basis report for approval of consultant. Single line diagram for electrical systems, water balance diagram for water related systems are part of basic engineering.

- **Sub Vendor Finalization**

Based on basic engineering approvals, tenders are floated for equipments and packages by the main contractor. Customer/ consultant participate in vendor finalisation.

- **Detail Engineering**

During this stage manufacturing and erection/ construction drawings are released.

- **Quality Plan approvals**

Before commencement of manufacturing/ construction Quality plan approval takes place. These plans detail out the stages of inspection and authority for clearing the product.

- **Logistics Finalization**

It is important to finalise the logistics for material movement from supplier works to project site. For imported consignments, the port of discharge is finalised. For road transport involving Over Dimension Consignments (ODC) the route is surveyed and approval is taken from rail/ road authorities.

- **Manufacturing and Supply**

Manufacture and supply commence after manufacturing drawing release.

- **Site Establishment**

Within days of award of work, process for site establishment starts. Facilities are made ready to receive and unload material. On readiness of stores dispatch clearance is given to suppliers.

- **Receipt of Material and Storage**

Material is received and stored for issue during construction.

■ **Construction**

Civil construction commences after establishing site. Foundations are made ready to suit receipt of equipment. Erection work starts after receipt of equipment.

■ **Pre-commissioning and Testing**

After completion of equipment erection, next stage is testing and pre-commissioning of individual equipment and systems.

■ **Commissioning**

After individual systems are commissioned, the whole plant as a single entity is commissioned.

■ **Trial Run and PG Test**

Trial run definition and period varies for different contracts. It is to establish the ability of the plant to operate at various loads continuously without any interruption. After successful trial run Performance Guarantee (PG) Test is conducted. Based on the PG Test the plant is taken over for commercial operation or corrections are made to achieve desired performance. Performance parameters are defined in contract.

■ **Contract Closing & Project Closure**

After successful completion of PG test, the plant is taken over by customer for commercial operation. Any deficiencies noticed are corrected before closing the contracts. Contract is closed when all pending issues are resolved, payment and bank guarantees returned. With this the project comes to closure.

2.2 Descriptive Questions

1. Explain the various stages involved in a project from concept to completion.

2. How an internal project does vary from infrastructure project?

3. What is brainstorming? How does it help in a project?

2.3 Multiple Choice Questions

1. For an Infrastructure project, Environmental Impact Analysis is

   i. Mandatory

   ii. Not Required

2. Financial closure of a project means

   i. All approvals are in place

   ii. All deliverables are defined

   iii. Cost of the project is finalised

   iv. Financial Institute have agreed to the Project cost and loan disbursement

   v. All of the above

   vi. None of the above

State True or False

SL	Statement	True	False
01	Environmental Impact analysis is not required for infrastructure projects		
02	Internal approvals precede external approvals		
03	Selection of consultants depends on complexity of project		
04	Every project requires a dedicated team		

ESTIMATION AND COSTING

There is a difference between Price and Cost. Price is what we pay to get goods or services. Cost is the expenditure incurred to make the goods or provide the service. In practice, Price is cost plus profit excluding taxes and duties applicable for supplying/ providing the product/ service.

There are many methods of costing in vogue. Cost is divided as Fixed cost, Variable cost and Overheads. It can also be divided as Direct and Indirect cost.

In Construction Management, the Fixed cost is the cost incurred in establishing the infrastructure. If part of the infrastructure can be moved after completion of the job and is useful then, depreciation for the project duration would become fixed cost; otherwise entire cost is fixed cost.

One point that needs to be remembered is costing is based on historical data. When the parameters change, one has to be cautious while using historical data.

3.1   Estimation

Every job to be undertaken needs to be estimated for cost likely to be incurred. This is because every project is unique and there are uncertainties. The uncertainties crop up because at the time of submitting an offer, design is not completed or certain unforeseen issues come up.

We have ready figures for certain elements, for balance we need to estimate based on proven logic.

Cost can be ascertained for some elements by obtaining budgetary offers. Budgetary offers are normally higher than actual cost.

There are three types of estimates one has to prepare

1. Estimate for quoting for a job

2. Estimate for award of work

3. Working estimate based on price obtained from customer

Working estimate helps in budget and cost control. It assigns limits on expenditure under every head based on the contract price. Many companies treat working estimate as a confidential document and is not shared with most of the team.

Basically there is no difference between the first 2 estimates, but many forget to consider all aspects of estimation and expect a much lower cost from a vendor than reasonable.

It is a standard practice that, the work cannot be awarded if the price exceeds estimate. Government tenders normally publish the expected cost of work and vendors are required to quote below the cost.

The elements of estimation are explained with a view that manufacture and supply are not part of scope of work.

3.2   Elements of Estimation

▪ Scope of Work

Scope of work details the activities to be carried out for the project. Specification details the parameters of performance.

▪ Project Schedule

Project schedule has effect on the cost. For expediting certain activities additional resources are required to be mobilised. Better equipment may be required to be deployed at a higher cost to meet a compressed schedule. New recruitment may be required if the existing manpower cannot be spared in time for meeting the schedule.

▪ **Facilities provided by Customer**

Customer can provide water, power, accommodation, space and equipment etc. These can be free of cost or at concessional charges. If adequate storage is not provided by customer, additional storage space needs to be identified outside the project boundary. This will necessarily cost additional amount by way of rent and development charges.

In the absence of adequate capacity EOT crane, alternate methods have to be explored.

▪ **Terminal Points**

Terminal point defines the boundary for work. There are many terminal points defined in a contract; like construction power at 11kV at one point or Coal at Bunker inlet.

The terminal points help in working out the cost of work. Schedule of Terminal points is part of the signed contract.

▪ **Exclusions**

Exclusions define the basis for working out cost and need to be communicated and accepted by both parties.

▪ **Definition of Commissioning**

Every contract defines the term commissioning. It is important to understand the requirements and work out cost.

▪ **Bill of Material / Quantities**

Bill of material/ quantities is very important to work out the cost of erection/ construction. It gives quantum and specification of material.

▪ **Weight to be erected**

This gives the breakup of weights and weight of individual components that require special care.

▪ **Joints to be welded**

Where welding is involved, the breakup of joints into pressure parts, structural, carbon steel, alloy steel and stainless steel is essential. Radiography requirement also needs to be specified. It helps in working out the cost and number of welders for meeting the schedule.

- **Soil Investigation details**

In civil works, soil investigation details play an important role. This gives indications of strata below the surface as hard rock, soft rock etc and the depth of these strata. Excavation cost depends on soil investigation details. Even after obtaining contract, specific area borehole investigation is necessary to design foundations.

- **Supply**

As per scope of work the quantities need to be worked out for various material in different sizes, specifications etc. Based on current market rates and expected rates during project duration, rate is worked out and expenditure calculated.

- **Tools & Plant**

Based on product details, customer facilities, project schedule and layout of the plant Tools & Plant required for the project is identified. Capacities and durations are identified. Based on the requirement decision to Buy or Hire equipment is taken.

- **Depreciation**

Depreciation is calculated for the own equipment deployed at site. Temporary facilities created are charged off during the period of project i.e. these are fully depreciated during the contract period. Depreciation is calculated as per both Company act and Income tax act. For estimation purpose it is only as per Company act.

- **Hire Charges**

These are calculated for both equipment and transport vehicles taken on rent for the period these are deployed.

- **Manpower**

Based on the product and schedule, histogram of persons to be deployed is prepared. This takes into account, skills, designation and experience. Requirement of specialists including expats is also worked out and cost worked out for the project. This includes salary and allowances paid to the staff.

- **Consumables**

Consumables can be for activities, equipment or office/ labour colony establishments. These need to be worked out and cost estimated.

- **Insurance**

Based on scope of work insurance cost is worked out for the project duration.

- **Transport**

Freight charges for material procured for the project needs to be calculated. This is normally charged on Rupees/ MT/ KM. Toll charges and octroi if applicable needs to be added.

The employees need to be brought to site and taken back daily. The children of staff have to attend nearby schools. If staff quarters are away from market place, trips need to be made to purchase grocery and vegetables. Finance staff needs to visit bank, stores staff needs to visit transporter's office/ godown. For all these vehicles need to be hired and operated. The expenditure on this based on distance to be travelled is estimated.

- **Travelling Expenses**

Persons have to travel for expediting work, inspection at vendor location and also to attend meetings at different locations. The expenditure is calculated based on number of trips, distance, mode/ class of travel.

Similarly the expenses towards transfer in/ out of staff are calculated. It is based on personal travel plus baggage transportation cost.

- **Material Handling, Preservation charges**

Material received at site needs to be unloaded, inspected and stacked for easy removal for erection and also safe storing. Material handling expenses are calculated on Rupees per MT. For special consignments like boiler drum, turbines, generator and transformers, item wise specific rates are applicable and need to be taken into account.

Material received needs to be preserved as per storage instructions of supplier. Cost of preservatives can be part of consumable expenses, however application charges have to be calculated based on supply/ erection schedule.

▪ Contractor / Labour Payment

Most of the work is carried out by engaging contractors. These can be pure labour supply contracts or including tools for erection along with labour and supervisors. The scope of tools & plant deployed by contractor varies as per availability of the equipment with company awarding the work. Depending on the logistics, the expenditure is calculated module wise.

For erection, it would be rate per MT classified into pressure parts, structural, ducts, ESP etc. For piping the rate could be on per MT basis or on inch diameter / inch metre basis. For STG it is a complete package. For electrical / C&I, it would be panel erection, cabling, termination etc. Pressure part welding and radiography are paid separately on quantum. Based on the product details, the expenditure is calculated.

For civil the rates are calculated for excavation, PCC, RCC, reinforcement, formwork in sq metres, staging, backfilling etc. As the level of concrete pouring increases the rate also increases for additional resources required.

If maintenance is on contract basis, then the expenditure towards this is also calculated.

Cleaning services for offices, staff colony and labour colony are required. These are based on mandays employed plus consumables.

The total expenditure for above is calculated module wise and summed.

▪ Fuel & Energy charges

Equipments need fuel and energy. Based on estimated running hours of equipment, the expenditure is calculated. If customer provides free electricity, then that is taken into consideration. However at many places, though power is free, taxes on power need to be paid by contractor.

▪ Safety Expenses

Safe working is of prime importance. Every person working at site needs to be provided with required Personnel Protection Equipment (PPE). Gloves and safety shoes get damaged and need replacement. Many companies provide 1 pair of safety shoes every year, but at site this may not last that long. Gum boots and raincoats are required during monsoon. Safety jackets with fluorescent stripes are now provided to all to identify the workmen at site. Helmets are long lasting, but their straps need replacement.

Barricading, safety posters, signage and other safety related material including fire extinguishers is required during the project. All this is a heavy and unavoidable expenditure.

The expenditure for safety needs to be worked out properly.

▪ Security Services Expenses

Every project site needs security to keep men and material safe. There are many agencies specialised in providing security related services. In addition to providing security guards, it can also provide intelligence regarding any threat to the installation. Some agencies provide dogs with handlers if there is such a requirement.

Caution needs to be exercised in selecting the agencies. Many agencies recruit guards from surrounding villages who are ill equipped to handle situations and many times afraid of locals.

Depending on the site conditions, requirement needs to be worked out for the strength of security detail at site. Payment is based on number of guards, their salary and allowances as per legislation plus supervision charges. Expenditure is calculated based on this.

3.3   Site Establishment Charges

Depending on the size of project, location and scope of work, temporary infrastructure needs to be built for completing the work as per schedule. Expenditure to be worked out is detailed below.

▪ Size, Number and Specification for Office Accommodation

There are standards for per person office accommodation requirements. Based on manpower requirement, office space is worked out. The supervisors and engineers need to be near to the work spot for close supervision. Enough space needs to be provided for drawings, documents, records and manuals for easy maintenance and retrieval.

The practice earlier was to construct temporary buildings with brickwork. With introduction of fully furnished portable office cabins, it has become easier to set up offices at fast pace. Portable toilet blocks are also available. For larger office space, option of prefabricated office is available that can be erected fast.

Based on the manpower and number of offices required, cost is worked out.

▪ Size, Number and Specification for Stores

Storage instructions are given by manufacturer. Based on project schedule, supply schedule and erection schedule the storage area needs to be worked out. Some material needs to be stored in air-conditioned area. Paints, fuel oil, gas cylinder, cement etc have their peculiarities and need to be stored accordingly. Storage space requirement is divided into

- ✓ Closed storage with/ without crane facility

- ✓ Air conditioned storage

- ✓ Semi closed storage

- ✓ Cement storage / silos

- ✓ Open storage

- ✓ Stores office

Open storage needs to be compacted, graded and developed with internal roads and drains so that material can be moved during heavy rains also. Wooden or RCC sleepers/ supports for stacking the material need to be arranged.

Small machined components can be stored safely in containers that come in various sizes.

If customer is not having adequate space inside the plant, then necessary storage space needs to be arranged nearby on rent basis.

Based on the above cost is worked out.

▪ Radiography source pit and room

Use, handling, transportation and storage of radioactive substances is controlled by Atomic Energy Regulatory Board (AERB). For sites where radiography work is carried out for long duration, construction of source pit and room is mandatory. The design of the pit and room needs to be approved by AERB. AERB authorised persons only can handle the source and take radiograph. The location of the room should be such that lesser number of persons move around.

▪ Construction Water scheme

Project construction requires huge quantity of water and uninterrupted water supply needs to be ensured for success of project. Depending on the location of the project site and agreement with customer, adequate capacity reservoir needs to be constructed. Borewell drilling for construction is prohibited and many times borewell water is not suitable for construction. Normally Customer provides construction water at one point in the project site; further distribution is in the scope of contractor.

From reservoir the water can be supplied by laying pipe lines to different locations and creating local sumps for water storage. If pipe line laying is not feasible, tankers need to be arranged for supplying water to different work spots.

Construction water scheme needs to be prepared and submitted to customer for approval on award of work. Based on the scheme the cost needs to be worked out.

▪ Drinking Water scheme

Every person at site needs to be provided adequate quantity of safe drinking water. This is mandated by law also. Based on peak strength of workforce adequate arrangement for drinking water has to be made. The requirement varies as per season. The cost of pumping and distributing drinking water or arranging by tankers needs to be calculated and added to the estimate.

▪ Construction Power scheme

Construction needs power. Normally customer provides construction power at one point in site at agreed voltage. Further distribution is in the scope of contractor. Many times the construction supply is from rural feeders and unreliable. In that case DG sets need to be installed at various locations in site. The power consumption is worked out based on number of equipments used and their power rating. The power requirement is not constant and varies as per the construction schedule. It is necessary to prudently work out the power requirement and install requisite transformers/ DG sets. Electricity suppliers charge a minimum amount as demand charges based on rating of transformers. One can augment the demand as the power requirement increases during the course of construction and save money.

The construction power scheme needs to be submitted to customer/ consultant for approval. It also needs to be approved by local electrical inspectorate. Charging of electrical installation is done only after obtaining permission from local electrical inspectorate. Excise duty has to be paid on power generated by DG sets.

The cost of equipment, cabling and maintenance needs to be calculated.

> ■ **Labour Colony – Specification, Size and capacity**

The requirements for labour colony are clearly specified in Building and other construction workers act 1996. It was earlier practice to give a piece of land to labour contractor and he would construct hutment for his labour. Power at one point would be made available by customer / EPC contractor. Providing hygienic conditions in labour colony is important for the progress of work. Similarly providing safe drinking water and general water is essential; otherwise there is labour unrest and work suffers.

The location of labour colony needs to be selected carefully. It has to be near to the plant so that worker can reach site conveniently and also can go for lunch during lunch break. With more than 2000 persons residing in labour colony, disposal of sewage and waste water without affecting surrounding villages is a major challenge.

Container based portable residential modules are available now. These permit more than one storey construction reducing the need for more space. With scarcity of skilled labour, available labour needs to be retained by providing proper facilities like electricity and fans in the residential units. Separate kitchens need to be provided for every contractor/ sub contractor.

Depending on the scope of work and area available, decision needs to be taken on type of construction for the labour colony. Provision of adequate number of toilets, bathrooms and washing places has to be made. Facility for storing drinking and general water needs to be made. Based on these elements, work out cost.

> ■ **Security guards' colony**

The accommodation for security staff has to be away from labour colony to avoid close contact between the two. Based on number of guards, adequate residential facility for security staff needs to be constructed. The cost is worked out for the same.

> ■ **Temporary Roads and Drains**

Design of permanent roads and drains including culverts takes time to get finalised. Till such time there is a need to provide temporary roads and drains including culverts to enable smooth construction progress. Based on details cost to be worked out.

> ■ **Site illumination scheme**

Permanent plant illumination work is awarded late in the contract period. For initial working and during construction adequate area lighting is required for round the clock working. Temporary roads and approaches need to be illuminated for safe walking. For proper security of material and site, the site needs to be illuminated to avoid thefts and security patrolling.

High mast towers provide general illumination for the site. Number and locations of high masts need to be fixed. For individual work spot lighting has to be arranged based on requirement. During structural erection, workers and welder should be able to climb the structure during night also. Accordingly lighting has to be arranged. Small DG set based portable lighting trolleys are also available for remote areas.

Based on the above scheme is prepared and submitted to customer/ consultant and agreed. Cost is calculated on the basis of material involved plus erection cost.

▪ Security office and Cabins

Office for security and safety staff needs to be constructed near main gate. This accommodates the staff and also has a training centre for imparting safety training to workers, staff and visitors. Gate pass issue and control is exercised from this office. Gate passes are issued to individuals and vehicles on permanent/ temporary basis.

At every entry/ exit point security cabin is required for verifying entry of authorised persons only. This also prevents labour from leaving the site without proper authorization.

Security watch towers with proper approach and lighting needs to be constructed at strategic locations. Based on the numbers cost is worked out.

Total cost is worked out for the above.

▪ Boundary wall/ fencing

Based on the scope of work and site conditions, decision to construct boundary wall/ fencing is taken. The plant boundary wall construction is part of billing schedule and should not be confused with this head. Even with existence of plant boundary wall/ fencing, individual areas need to be separated for restricting access. Storage area needs to be fenced. In addition to this, transformer yards and fabrication yards are fenced. Offices also can be fenced to prevent physical threat to employees.

Cost has to be estimated based on above, including fabrication of gates for individual fenced areas.

▪ Area grading and levelling

Main plant area grading and levelling is part of the billing schedule. However non plant areas need to be graded and leveled for use during construction. This needs to be estimated based on quantum of work.

▪ **Weigh Bridge and control room**

Weigh Bridge is required during construction for weighing the consignments coming in. Based on the weight only, bills are passed. Weigh Bridge has to be of adequate capacity and length to accommodate the long trailers coming in site.

Weigh Bridge control room houses the computer, electronic hardware and ticket printer. This is air conditioned room with furniture for operator and hardware.

In the absence of Weigh Bridge in site, stores staff has to accompany the vehicles to public weigh bridges on highway for ascertaining weight of consignment. This is a time consuming affair and person has to be deputed for this work only.

Cost is worked out based on above.

▪ **Radiography Source pit and Room**

The drawing for this facility needs to be submitted to AERB and approved by AERB. The location has to be such that it is nearer to work spot but away from public movement.

▪ **On site Facilities for Labour as per statutory requirements**

These include

- ➢ Canteen(s) for labour at short distance from work spot. Many contractors operate their own canteens for their staff. Hygiene and area cleanliness is a major issue

- ➢ Creche for children of women workers

- ➢ Adequate number of urinals and toilets for men and women near work spots, with water supply and waste disposal arrangement

- ➢ Two wheeler stand(s) for parking vehicles

- ➢ Shelter near work spots for summer and rainy seasons

- ➢ Safe drinking water

- ➢ Primary Health centre with paramedics and ambulance

Estimate should take into account all the above elements

3.4 Information Technology Infrastructure

This is an essential part of any business activity. Earlier there was dependence of post and couriers for timely receipt of drawings and documents. The current trend is to receive drawings and documents through internet. Every site requires strong IT infrastructure for smooth flow of work. Many sites are now connected to ERP systems at Head Office for online transactions and effective control. This includes

- ✓ Internet connection – Enough bandwidth is required for the net connection for fast downloading of drawings files and video conferencing. Suitable alternatives need to be explored.

- ✓ Routers and modems – These are required for connecting the internet connection with users at different locations within the site. Many routers are available with firewall capabilities.

- ✓ Servers – File servers, antivirus servers, print servers, VPN server and DNS servers are required. Depending on the requirement, it can be integrated into 1 or 2 servers.

- ✓ Desktop computers – As per requirement

- ✓ Work stations – Engineering workstations are required for preparing or modifying CAD drawings. Design modifications can be marked on the drawing and sent for approval. Similarly part of the drawing can be printed for attaching with work permits. Underground utilities can be marked on plot plan with these. As built drawing preparation also becomes easy with these.

- ✓ Laptops – As per requirement and company policy.

- ✓ A0 size plotters – For a large site minimum 2 plotters are required for taking print outs of drawings.

- ✓ Printers – A3/ A4 size printers are required in different sections and location at site.

- ✓ Scanners – A3/ A4 size scanners are required in various departments within site for scanning various documents.

- ✓ Photocopiers – A3/ A4 size photocopiers are required in various departments.

- ✓ Projection systems – For use in conference room and training centre.

✓ Video conferencing equipment – Many meetings are held using video conferencing. This consumes bandwidth to a large extent.

✓ Local area network within the site – Hardwired or Wi-Fi based network is required for connecting computers.

The requirement varies as per the scope of work, number of locations within the site and number of persons accessing the network.

Based on above the expenditure both one time (Capital) and recurring (Revenue) needs to be calculated.

3.5   Sundry Expenses

These include a variety of various different small costs that are classified as a whole and are thus not assigned to any one particular specific ledger account.

o   Printing, Stationery, Postage

The requirement is assumed based on past experience of similar size sites.

o   Bank Charges

These depend on the relationship between company and dealing banks. With NEFT and RTGS writing cheques has come down but still these charges are incurred and need to be estimated.

o   Electricity, Water charges

Electricity and water bills charges for the establishment and residences is calculated here.

o   Rent residential

This takes into account the rents for houses taken for accommodating staff.

o   Taxi/ Hired vehicles for manpower

Buses, mini buses, jeeps and cars are taken on rent for transporting staff and their families. This accounts the expenditure.

o   Guest entertainment expenses

Many occasions are celebrated during the project execution. Persons are treated as company guest and entertained. Expenditure towards this estimated.

o   Guest House/ Hotel expenses

Arrangement for visiting officials and employees on transfer are made for few days on arriving at site. Guest house needs to be maintained with proper kitchen and furnishing. Expenditure towards this is estimated and accounted.

3.6   Summing up

o   The entire above estimate is totalled to derive basic cost.

o   Overheads are taken as a percentage based on individual company policy and applied on basic cost.

o   Contingency @ 2% - The accuracy of estimate cannot be 100% and hence 2% of basic cost is taken as contingency for unforeseen expenditure.

o   Profit 10/9 % - To get a profit of 10% on Rupees 100, cost is 90. Hence profit is calculated as 10/9 % of cost. If expected profit margin is more accordingly the formula needs to be changed.

o   Taxes and Duties to be computed – Custom duty, Excise duty, Sales Tax/ VAT, Service Tax & cess on it and Work contract tax need to be calculated on the basic cost. Most of the contract price is exclusive of taxes and duties, but the quantum needs to be indicated. Even if the contract price is inclusive of taxes and duties, this figure needs to be shared with customer. This helps in resolving issues if the tax structure changes during the contract period.

o   Negotiation margin as per company policy – Every buyer likes to negotiate and no sale is complete without a discount. For this negotiation margin is kept within which marketing can offer discount.

3.7   Descriptive Questions

1. Describe the process of Estimation for a Supply and Supervision contract. How does it differ from Supply and Construction contract

2. What is Site infrastructure? Explain various components of site infrastructure

3. What are different components of estimation?

3.8   Multiple Choice Questions

1. Budgetary offers are collected to

    i.   To create a data bank
    ii. To determine probable cost
    iii. To promote awareness about company
    iv. All of the above

State True or False

SL	Statement	True	False
01	Defining Terminal Points is a must		
02	Exclusions define cost not accounted		
03	Budgetary offers are a good tool in estimation		
04	Scope of work should be unambiguous		
05	Support of other departments is required in estimation		

MATERIAL MANAGEMENT

It covers the acquisition of material and replacements, quality control of purchasing and ordering and the standards involved in ordering, shipping, and warehousing.

Material Management function covers

- **Requirement processing**
- **Vendor Management**
- **Commercial & Contract Management**
- **Procurement**
- **Assistance in Quality Assurance & Control**
- **Insurance**
- **Stores**
- **Document Control**

The processes are described below.

- ➢ Requirement processing involves
  - ✓ Receiving Indent

Indent is an instrument conveying the requirement of product/ processes / service to be procured to Purchase/ Contracting department. It contains Purchasing data.

Purchasing data for product/ processes/ services contains following details as applicable:

- ✓ Title or other positive identification

- ✓ Applicable Indian/International standards

- ✓ Capacity, Performance rating, and other technical requirements

- ✓ Applicable drawings and data sheet

- ✓ Statutory and Regulatory requirements including Material Safety Data sheet (MSDS)

- ✓ Quality Management System requirements

- ✓ Requirements for approval of Product, Procedures, Processes, and Personnel qualification

- ✓ Special requirements regarding, inspection, testing, staggering of delivery, guarantee, packing, insurance, and consignee details etc. as applicable.

- ✓ Budget provision for the year and for the contract.

The above purchasing data becomes part of Indent.

In every organisation the responsibility of raising of an Indent is defined for various products and services the organisation utilizes. The department concern raises an Indent and sends it to Purchase/ Material Planning department.

- ✓ Checking for correctness and completeness of the Indent

Purchase/ Material Planning department person will check the Indent for correctness and completeness. Any ambiguity in specification would be removed after discussing with Indentor.

- ✓ Checking for compatibility with Contract requirements

Purchase/ Material Planning department will check the compatibility with contract requirements by discussing it with Contract Management/ Project Management. Any discrepancy would be removed at this stage.

✓ Register Indent and Communicate

On fulfilling the above checks, the Indent would be assigned a registration number and Indentor's copy would be sent to him with this registration number. In all further correspondence this registration number is referred.

✓ Types of Indent

Indents can be for purchase against a customer contract or for internal consumption. Indents are classified as

- Normal
- Emergency
- Proprietary

In case of Normal Indent, the normal time cycle for the purchase process is followed. For emergency indents quotes may be collected by phone or by visiting the vendors in person by a team. For proprietary indent the particular vendor only is contacted for getting offer.

✓ Floating enquiries for Purchase of Product and Services as per schedule

Based on number of parties invited to bid, Enquiries can be

- Single Tender or Nomination basis
- Limited Tender
- Open Tender

In case of emergency requirement, Single Tender can be followed by recording the reasons for the action with approval of competent authority.

In case of Limited Tender, few selected parties from Vendor list are given the tender document. In this case also the action needs to be supported by valid reasoning for omitting other probable vendors.

When the Tender document is available to any vendor by way of sale of the document, it is Open Tender. The intimation to vendors is given by either advertising in News papers or publishing on web. Most companies have a section reserved on their web site for Tenders in offing. If all the vendors on the approved Vendor list are given opportunity to quote, then also the tender is presumed to be open tender.

The bidding process can involve either domestic bidders only or allow participation of bidders from outside the country. Based on this it is called either Domestic competitive bid or International/ Global competitive bid.

For small value items, many times local purchase is authorised. In this case a committee visits local market and obtains quotations and decides on placing Purchase order. In Government and Government controlled companies, there is a requirement of minimum three quotations for placing order.

The time cycle for placing Purchase order from the date of enquiry floating depends on the process followed.

✓ Receipt of quotes

Most of the Tender / Bid consist of following

1. Security Deposit

2. General Terms and Conditions

3. Technical Specification and conditions

4. Special conditions

5. Price Bid

Every organisation has standard General Terms and Conditions that are approved by management. These need to be attached with every tender. Any change in these conditions, needs prior approval of management. These are to take care of legal aspects of the order and also describe the payment terms. The commercial terms and guarantee requirement are specified in this section.

Security deposit is obtained from all bidders to ensure that only interested parties participate in the bidding process. Though it is in proportion to cost of purchase, there is upper limit fixed by every organisation and that is the amount of security deposit collected for costly purchases. Till 1990s Fixed Deposit Receipts, National Savings Certificates etc was accepted as Security deposit; but it is not acceptable now.

For regular vendors, acceptance of General Terms and Conditions is obtained as a onetime exercise. This eliminates the need to attach bulky document with every tender. Similarly a bank guarantee of sufficient amount on yearly basis is taken from a regular vendor that covers most of the jobs they quote. This also eliminates the process of receiving security deposit, depositing the amount in bank and then returning it to bidders other than successful bidder.

Technical specification and conditions describes the technical requirement for the product or service. It also enumerates various standards to be followed.

Special conditions section describes any site/ contract specific requirement. This includes quality, safety and other requirements. Inspection by a customer in whose premises the product is to be installed or service to be performed or agency appointed by customer are described in this section. It also describes expected relationship with other contractors at site.

Price Bid gives the format, in which price has to be quoted.

Though all the five parts can be received together in one envelope, the practice is to receive the parts in 4 envelopes as follows

1. Security Deposit

2. General Terms and Conditions

3. Technical specifications and conditions; Special Conditions

4. Price Bid

If all parts of the bid are in same envelope, it is a composite bid. If separate envelopes are required for every part of the bid, then it is called multipart bidding system. All the envelopes are not opened at the same time.

The enquiry specifies date and time by which the offers are to be delivered to a specified person and address. It also specifies the date, time and venue for opening the bids. Representatives of bidders can be present when the bids are opened. Late offers and incomplete offers can get summarily rejected.

✓ Review of offers and resolution of issues

On specified date and time all the bids received are opened in the presence of a committee appointed for the purpose. Signatures of all present for tender opening are taken on a tender opening register. This includes all the representatives of bidders.

If the bidding is multipart bidding, then on the tender opening date it is verified that requisite number of envelopes are submitted by all bidders. Validity and correctness of security deposit is checked. In case of discrepancy the bid is rejected.

One important aspect is availability of Power of Attorney in the name of person who has signed the offer from bidder, if he is not owner/ partner or director of the company participating in the bid.

Multipart bidding system is preferred because price alone does not become deciding criteria. In composite bid system, there will always be pressure to accept the lowest bid though technically the bid may not be meeting all the criteria.

In Multipart bidding system, first the envelope containing General Terms and Conditions is opened for all valid bids. On review of the bids, the deviations sought by bidders are recorded for further action. Some of the deviations may be acceptable while other deviations require discussion to arrive at a solution. Acceptable deviations are communicated in writing to all bidders. Meetings are conducted with individual bidders for resolving other deviations. On discussion certain deviations may be accepted by purchaser. Bidder is asked to withdraw balance deviations, out of which he may agree to withdraw certain deviations, but insist on keeping other deviations. In that case the bidder is informed that cost of deviation would be loaded on his price. This excludes certain basic requirements on which no deviation is acceptable. Offer with such deviations would get rejected at this stage and the bidder informed.

The envelope of technical offer is opened at this stage and reviewed for completeness and details of offer. Missing details are sought from bidders and queries raised by them answered. Discussions held with individual bidders and decision on Technical deviations sought is communicated. Technical deviations that are accepted are communicated to all eligible bidders. Bidders are informed to withdraw balance deviations. If a Bidder insists on retaining a deviation, bidder is informed that his bid will be loaded for cost of deviation.

At this stage list of technically and commercially acceptable bidders is prepared. Based on the discussions and acceptance of deviations, bidders are requested to submit their revised price bid if necessary.

The Price bids are opened on a specified date in presence of the eligible bidders. Based on this a comparative statement is prepared to decide the lowest bid. Negotiations are held with bidders to explore the possibility of further reduction in price and bring it nearer to expected price. In Government and Public sector negotiation can be held only with the lowest bidder as per Central Vigilance Commission (CVC) guidelines.

   ✓   Placement of Order as per schedule

Based on comparative statement and negotiations, a purchase proposal is made and Letter of Intent (LOI) issued. After acceptance of LOI by bidder and receipt of Bank guarantee, the detailed Purchase / Work order is released. At this stage the security deposit of all other bidders is returned.

➢ Vendor Management

For Vendor/Subcontractor selection, their credentials are reviewed by a Committee consisting of representatives of purchase, finance, and concerned departments as authorised by Management.

Vendors/Subcontractors are assessed and enlisted based on following criteria as applicable

- Past performance

- Feedback from other users

- Technical and Financial capability

- Product certification by independent certifying authorities.

- Ability to provide adequate resources

- Ability to meet quality requirement

- Conformance to statutory and regulatory requirements, as applicable

Selected Vendors/Subcontractors are registered and list is maintained by Purchase/ Contracting department.

List of selected Vendors/Subcontractors needs to be updated periodically based on performance evaluation and re-evaluation by the committee.

➢ Coordination for changes in Customer's approved vendor list and introduction of new vendors to customer/ consultant

Every customer contract lists the acceptable vendors for various products/ packages / services. These are based on the experience and preference of customer/ consultant. Many vendors who are part of company's approved vendor list do not find a place in this customer's list. The number of probable vendors is also less. In such case, it is necessary to add additional vendors to customer's list to generate competition and get a suitable price. For this, credentials of these vendors need to be provided to Customer/ Consultant to obtain their approval. This process will continue till all items/ services are procured for a project.

➢ Commercial & Contract Management in Procurement

Finalising the billing schedule is part of this process. The progress of the supply/ work needs to be monitored. Receiving dispatch documents, bills and scrutinising them for correctness is required. Bills are processed for payment based on certification from stores for receipt & acceptance of material and certification from execution for the quantum of work completed.

Any disputes need to be resolved and contract closed.

> ➢ Assistance in Quality Assurance & Control

This involves receiving Quality plans from vendors and forwarding them to Quality Assurance for approval. Arranging meetings for resolving any quality related issues between vendor and QA &C.

At stores Quality control is involved for receipt inspection and audit of preservation of material.

> ➢ Insurance

Insurance is the equitable transfer of the risk of a loss, from one entity to another in exchange for payment. It is a form of risk management primarily used to hedge against the risk of a contingent, uncertain loss. An insurer, or insurance carrier, is a company selling the insurance; the insured, or policyholder, is the person or entity buying the insurance policy. The amount of money to be charged for a certain amount of insurance coverage is called the premium.

Insurance contracts are generally considered contracts of adhesion because the insurer draws up the contract and the insured has little or no ability to make material changes to it. This is interpreted to mean that the insurer bears the burden if there is any ambiguity in any terms of the contract. Insurance policies are sold without the policyholder even seeing a copy of the contract.

Insurance contracts are **aleatory** in that the amounts exchanged by the insured and insurer are unequal and depend upon uncertain future events.

In infrastructure projects following types of insurance is involved

1. Project Insurance

This can be taken as a whole are divided into

- Marine – For transit damages from supplier works to project site.

- Storage cum Erection – For loss during storage and Erection

2. Public Liability

   This is to cater to liability accruing from an accident where surrounding population is affected.

3. Motor vehicle

   This insurance protects the policyholder against financial loss in the event of an incident involving a vehicle they own.

Coverage typically includes:

- Property coverage, for damage to or theft of the vehicle

- Liability coverage, for the legal responsibility to others for bodily injury or property damage

- Medical coverage, for the cost of treating injuries, rehabilitation and sometimes lost wages and funeral expenses

4. Employees group insurance

   This covers the employees of an organisation. Names and age of employees need to be given to insurance company. The changes in the list are communicated once in a year to insurance company. In the event of death of an employee his family is paid the sum assured. (The premium for policy is then individual policy.)

5. Workmen compensation

   Insurance needs to be taken for all workers either under workmen compensation policy or Employees State Insurance scheme. The details are inspected by Labour department.

6. Equipment insurance

   This policy covers cost associated with breakdown and maintenance of equipment.

7. Third party damage

   In every project multiple agencies work together. This policy covers any accidental damage caused to other contractor's property or work. It also covers damages to facilities provided by customer.

The insurance cover for the project is taken either by the owner or EPC contractor as per contract. For this a detailed break up of equipment to be covered with cost needs to be given for calculating the sum assured. Based on the sum assured and tenure, premium is calculated. The premiums are paid every quarter. The sum assured being huge, normally the cover is given by a consortium of insurance companies with one agency as lead agency entering into contract. It is customary for the insurance companies to seek reinsurance globally to cover their overall exposure.

4.1   Franchise deductible / Excess

A minimum amount of loss that must be incurred before insurance coverage applies. A franchise deductible differs from an ordinary deductible in that, once it is met; the entire amount of the loss is paid, subject to the policy limit. Franchise deductibles can be stated either as an amount or as a percentage of the policy limit.

This has effect on the premium amount. Higher the deductible / excess, lower is the premium.

This deductible / excess have effect on claim management process also. Any repair/ replacement lesser than deductible has to be borne by insured and claim is not registered by underwriters.

4.2   Insurance claim management

Every insurance company has its own claim settlement procedure and formats. Persons dealing with insurance claim should get acquainted with the procedure and get the formats printed for use.

The basics of insurance claim management are

1. To Maintain Insurance Claim Register – This is audited by Internal audit team / Dept. of company. It helps in keeping track of all claims and their current status. It also helps in finding the claim amount and settlement amount.

2. To intimate underwriters of dispatch of goods from supplier on the same day. In case of transit loss/ damage, if this intimation is not given the claim would not be entertained.

3. On noticing damage/ loss in transit, record it on transporter's copy of LR and raise a demand on transporter with a copy to underwriters. Take a photograph of the damages to support the claim.

4. In case of theft from project site, lodge First Information Report (FIR) with a local police station and get a copy. Intimate the underwriters.

5. In case of damage during handling / erection, get an incident report from the executive concerned how the damage took place. Take photographs of the damages and the location where the incident took place. Intimate underwriters. Assess the losses and in case of replacement, get cost of replacement from supplier.

6. In case of vehicle accident on public roads, FIR is essential. Cost of repair to be obtained from authorised service centre. Intimate the underwriters.

7. Time is of essence and policy defines the time limit for intimation for every type of claim.

8. Based on the extent of loss and type of equipment involved, insurance company deputes appropriate surveyor.

9. Insurance claim is passed by insurance company on the basis of surveyor's report.

10. Insurance is on the basis of mutual trust and never give impression that organisation is trying to derive undue benefit.

➢ Stores

✓ Receipt, storage, preservation and issue of material

To effectively manage stores the following are required

- Receive List of Work/ Purchase Orders

- Receive copies of Work/ Purchase Orders

- Receive Shipping List/ Bill of Material for the above

- Receive Storage and Preservation manuals from the manufacturer

- Arrange racks and bins for components to be stored in closed sheds.

- Arrange wooden/ RCC sleepers, tarpaulins and polythene sheets for keeping material safe in open storage yard.

- Arrange for material handling and lifting equipment for unloading/ loading of material.

- Arrange IT facilities and internet connection for communication and updating records. A scanner/ photocopier and printer are essential for stores.

- Arrange staff and labour for various activities.

- Prepare a material stock register

✓ Receipt of Material

- Maintain LR, RR and Day Book Registers

Lorry Receipt (LR) and Railway Receipt (RR) are the dispatch documents sent by supplier to consignee. In most of the cases the material is consigned to the customer and material is received/ stored/ issued by contractor on his behalf.

Every LR and RR that is received at site is entered in LR and RR registers. LR and RR register help in keeping track of material in transit. It also helps in accounting freight charges.

Day Book (DB) register is the register in which every delivery challan and consignment received at store is entered. It also gives the location where the material is unloaded before taking into stock. It is common practice to have multiple DB registers in the store, with separate DB registers for major suppliers. Separate DB registers are maintained for T&P and local purchase. This helps in reconciling the supplies.

- Receipt of Material after entry at Gate

The material coming to site has different purposes of entering the site as follows

1. Material for construction billed on customer

2. Material for enabling construction that belongs to contractor that is taken back after work completion. This includes structural steel, scaffolds etc.

3. Tools & Plant that is used during construction and moves out after work completion.

4. Gas cylinders, office furniture and equipment and other things belonging to contractors and to be taken out after work completion.

As can be seen, some material gets consumed at site while balance goes out. Hence there is a need to make an entry at the security gate register of such returnable material.

After entry of material at plant security gate, the consignment is brought to the store for unloading. Store keeper would tell the driver where to take the material and gets it unloaded.

In case of sand and stone aggregate, QC is involved to check the quality of material before unloading the material. Material is unloaded only if QC clears the consignment. For this purpose a dedicated civil QC person needs to be stationed at stores in shifts, as the transporter is in a hurry to unload to make another trip. Unsatisfactory consignment is returned without unloading.

Before unloading the material it is ascertained that it belongs to the project. If not, then it is not unloaded and transporter is asked to it back after informing the consignor.

It is also checked whether there is any visible damage to the material or boxes. If any damage is noticed, then entry is made on the transporter's copy of LR for claiming damages.

✓   Receipt Inspection

Material coming in stores needs to be inspected for correctness before taking into stock and releasing payment for supply. The steps involved are

*   Check whether material belongs to the Project

It is customary for manufacturers to load more than one project's material on a single truck/ trailer. The material is unloaded at one project and then balance material is taken to next project. Sometimes there can be mix up also. Hence it is necessary to check that the material received belongs to the project before unloading.

*   Involve QC and Execution group to inspect the material

Stores personnel need not be knowledgeable in most of the products received. It is therefore essential to associate QC and Execution group to inspect the material before taking into stock.

- Check for royalty payment receipts for sand and aggregates.

State governments levy royalty on sand and aggregate and it has to be paid by supplier. Vehicles can be stopped on road and seized if royalty receipt is not showed.

- Check for Material Safety Data Sheet (MSDS) for materials that require this to accompany consignment.

The MSDS is a detailed informational document prepared by the manufacturer or importer of a hazardous chemical. It describes the physical and chemical properties of the product. MSDS's contain useful information such as flash point, toxicity, procedures for spills and leaks, and storage guidelines. Information included in a Material Safety Data Sheet aids in the selection of safe products, helps you understand the potential health and physical hazards of a chemical and describes how to respond effectively to exposure situations.

- Check for damages and correctness of Quantity

During transportation some damages do take place. This can happen if the material is not properly packed and slings touch material directly. If any visible damage exists take a photograph. Write on LR copy and take signature of driver for making claim on transporter.

Though the Bill of material/ shipping list indicates quantity in numbers, it is customary for sending small items like fasteners, hooks etc by weight taking into account average weight of components. It is essential to check the quantity and raise short supply note to supplier if necessary.

- In case of problem raise nonconformity/ claim

In case of damages in transit, Insurance policy expects receiver to notify the transporter about the damages and lodge a claim. Insurance policy specifies time limit for verification and this needs to be adhered to.

- Handover O&M spares received to customer

In some cases O&M spares are also ordered along with main supply. These need to be handed over to customer for safe keeping and acknowledgement obtained for records.

✓ Take into Stock

After unloading and receipt inspection, the material is taken into stock. Entries are made against corresponding item, the quantity received and its location in stores.

✓ Issue of Material

Issue of material can be chargeable or free as per contract terms. The issue can be returnable or non returnable. Returnable issue voucher is for tools & plant which are returned after use. Construction material is issued as non returnable as it is to be consumed. As there is cost involved, material issue needs to be authorised by designated persons. Different types of issue vouchers are used for above categories. Care has to be taken while issuing material as there are many similar looking components but with different characteristics.

After receiving issue voucher at stores, contractor is permitted to lift the material and a gate pass is made for allowing the material to go out of stores area by security guard. After this entry is made in stock register and stock is reduced.

✓ Receipt of returns and their segregation

There are many types of returns coming to stores

- Empty cans of paints and chemicals
- Empty oil barrels
- Empty gas cylinders
- Empty shipping containers
- Scrap of wooden packing material
- Used batteries
- Excess material remaining after completion of erection – In shipping list certain material is issued in excess of drawing requirement for possible damage or loss during erection. Such excess material is returned to stores.
- Cut pieces of raw material after fabrication completion
- Metal scrap after fabrication completion. This also includes damaged parts.

o   Tools & plant and fixtures after use

All the above returns need to be inspected.

Equipment needs to be checked for proper operation. If discrepancy is noticed, equipment needs to be repaired and repair/ replacement charges levied on contractor/ user.

✓   Disposal of scrap and rejects

Cut bits of metal are screened to ensure that there no usable pieces. Scrap and unusable cut bits are stored separately and auctioned. Metal scrap of damaged parts is also auctioned after insurance claim is settled.

Empty cans and barrels are also auctioned.

Used batteries can be given to battery dealer only and cannot be auctioned.

✓   Follow up with transporters for delivery/ transport of material

Most of the material comes to site by dedicated trucks, trailers. However some material is dispatched as smalls and comes to transporter's godown. These consignments need to be collected from godown. In some cases the original dispatch documents are sent through bank and cleared only when bank gets payment for the documents. Small consignments sent by railway also need to be collected from destination railway station. Stores group has to keep track of these consignments and collect them without attracting demurrage. If some consignment is not received within specific time, then stores group need to check up with transporter for the whereabouts of the consignment and take necessary action.

✓   Procurement/ Issue of consumables

Major material is supplied by manufacturers and other supplies are organized from company HO. However consumables are normally procured from site only in small quantities to avoid wastage. These are taken in stock and then issued. Sometimes in emergency other material also is procured from site after following due procedure.

✓   Material reconciliation with customer

Material from manufacturers is consigned to customer and it is held in trust by EPC contractor. Part of the supply payment is released only after certification of receipt by site stores. It is also essential to ensure that contractor claims erection/ construction of material received only. For this regular reconciliation is required with customer for expediting bill processing.

✓ Document Control

Material management documents serve as financial instruments and can be asked to be produced in court of law. As such making them legible and preserving them for a period prescribed by law is essential. It is a practice to preserve the documents for a period of seven years from the date of closing a contract.

▪ Challenges faced by Materials Managers

✓ The major challenge that material managers face is maintaining a consistent flow of material for production. There are many factors that inhibit the accuracy of inventory which results in production shortages, premium freight, and often inventory adjustments.

✓ The major issues that all material managers face are incorrect bills of material, inaccurate cycle counts, un-reported scrap, shipping errors, receiving errors, and production reporting errors.

✓ At construction sites, problems come with

- o Sand and aggregate – The receipt and payment is normally by volume but issue is by weight for concrete mixers. There is tendency to divert sand and aggregate for unplanned activities without informing the concerned. There is no issue voucher created for this consumption and the balance stock does not match.

- o Cement – Batching plant output figures do not match with pour card quantities.

- o Bricks – Bricks do get damaged during handling and some quantity is taken without permission for other use.

- o Steel cut bits – All cut bits are not collected and returned to stores. Some cut bits are used for fabricating smaller components or fit up pieces. Normally fabrication contractor takes receipt from other users for these cut bits.

- o Welding electrodes – Consumption exceeds design quantity released. TIG wires are not supplied in enough quantity.

o   Fasteners – Loss during erection and thread damages noticed during tightening. Nominal weight not matching with expected quantity.

For most of the above, contractors are penalised, but the issue needs to be sorted out.

✓   Material managers have striven to determine how to manage these issues since the beginning of the industrial revolution. Although there are no known methods that eliminate the above mentioned inventory accuracy inhibitors, there are best methods available to eliminate the impact upon maintaining an uninterrupted flow of material for production.

4.3   Descriptive Questions

1.   Describe the various processes involved in Material Management

2.   Describe the steps in a International Competitive bid finalisation upto order placement

3.   Describe the processes in Stores from Receipt to Issue and disposal.

4.   What are different types of insurance policies?

5.   What is Indent? What requirements it needs to fulfil?

4.4   Multiple Choice Questions

1.   A Tender is called Open Tender when

    i.   It is published in News Paper

    ii.   It is published on Internet

    iii.   All qualified parties in Vendor List receive the Tender

    iv.   All of the above

    v.   None of the above

2.   Receipt Inspection is carried out to verify

    i.   Conformance to specification in PO

    ii.   Correctness of quantity as per despatch documents

    iii.   Physical damages

    iv.   All of the above

    v.   None of the above

3.   Used Batteries can be disposed off

    i.   In any way

    ii.   To a scrap dealer

    iii.   To a Battery manufacturer or Battery dealer

4.   Budgetary offers are collected to

    i.   To create a data bank

    ii.   To determine probable cost

    iii.   To promote awareness about company

    iv.   All of the above

5.   Liquidated Damages can be claimed to

    i.   Any extent

    ii.   To the extent contract provides for

    iii.   To the extent of actual/ estimated loss limited to contract provision

6.   For construction cement is issued on

    i.   Last in First out

    ii.   First in Last out

    iii.   First in First out

    iv.   Batch wise First in First out

State True or False

SL	Statement	True	False
01	Receipt Inspection is not required		
02	Used batteries from plant can be sold as scrap to a scrap dealer		
03	Reducing bank guarantee value is treated as productivity improvement		
04	Inventory levels have no impact on Productivity of an organisation		
05	Two part bidding system is preferred		
06	Vendor registration is not necessary		
07	Negotiation is must for obtaining right price		

LOGISTICS IN A PROJECT

- The *Oxford English Dictionary* defines logistics as "the branch of military science relating to procuring, maintaining and transporting material, personnel and facilities.

- Another dictionary definition is "the time-related positioning of resources". As such, logistics is commonly seen as a branch of engineering that creates "people systems" rather than "machine systems".

- According to the Council of Logistics Management, logistics includes the integrated planning, control, realization, and monitoring of all internal and network-wide material, part, and product flow, including the necessary information flow, industrial and trading companies along the complete value-added chain (and product life cycle) for the purpose of conforming to customer requirements.

- Logistics is the process of planning, implementing, and controlling the effective and efficient flow of goods and services from the point of origin to the point of consumption.

5.1   Logistics – Process

Though this process belongs to a supply contract, a construction management professional should be aware of the intricacies involved. This also is applicable for transporting heavy construction equipment belonging to Construction Company.

> ▪ **Define the point of Consumption (Project site)**

For a construction site, point of consumption is the project site. However if heavy fabrication is ordered on fabricators outside the site locality, then one has to consider transportation of raw steel to these suppliers.

> ▪ **List all points of origin (Suppliers)**

All the suppliers, whether indigenous or foreign, become points of origin.

> ▪ **List equipment that requires special care in transportation and their details (Weight, size etc)**

Normally highways are designed for a maximum axle load of 10 MT and the Motor Vehicle act prohibits size of consignment larger than the truck body. Height restrictions are also in force. Any consignment that exceeds these limits is considered Over Dimension Consignment (ODC). For ODC permission needs to be taken from highway authorities and there is restriction on movement timing.

All the rail tracks also are not designed for movement of heavy loads. Specific permission from Railway Safety Commissioner is required for a route and the consignment can travel on that route after safety inspection.

> ▪ **Identify vehicles suitable for transportation**

Depending on the size and dimensions of consignment, suitable vehicles need to be identified. Multi axle vehicles with hydraulic control are available for heavy consignment requiring lesser radius of curvature at turnings.

> ▪ **Define route for transportation of above and check suitability for the load to be carried**

Some roads are single lane; others double lane or having multiple lanes. The road width and load carrying capacity vary accordingly. Bridges also have different load carrying capacities. For normal loads to be transported the transporters have distance charts and charge freight in Rupees per MT per KM based on these charts.

For ODC special survey needs to be undertaken of alternate routes for speedy delivery. This covers possible routes, width restrictions, height restrictions and load restrictions on some section as well as bridges.

• **Define actions to be taken and approvals needed**

In case of railway transport, the approval is from Railway Safety Commissioner. So it is one point approval but takes time.

In case of road transport, one has to take into account the number of states to be crossed, probable roads, and actions to be taken for old bridges. Where river water levels are low, constructing a temporary culvert may be feasible, in other cases, laying of steel plates to distribute point load is a solution. However after laying steel plates also the load may exceed permissible load. In such case that route cannot be used.

All highways do not have Rail Over Bridges (ROB) and the consignment may have to cross railway line. With railway electrification, there may be requirement of switching off power supply during transit of the consignment at railway crossing. For this, permission from concerned railway division is required, as only they can switch off power supply for limited period. The timing for this transit needs to be approved in advance.

If the consignment passes through villages, some houses may get damaged or need to be demolished during movement. Compensation needs to be fixed in advance and paid. Permission from village Panchayat also needs to be taken.

In some cases, if possible the movement of cargo by ship is also considered if feasible. Shipment may reduce transit time and also number of states involved in transit.

• **Get approval of client, underwriters and authorities controlling the route**

The underwriters play a major role in approval of the proposed route as they are bearing the risk. Approval of client and underwriters is taken for proposed route. Once the route is finalised, approval from highway authorities both central and state is taken.

• **Engage suitable transporter with proven capability**

All transporters do not have suitable vehicles for transporting the heavy consignment. As such suitable transporter with proven capability needs to be selected.

• **Alert local authorities when the consignment is about to pass their jurisdiction**

This is essential to avoid any disturbance while the vehicle moves on the route. The local authorities also may restrict movement in specified duration to avoid public nuisance.

> ▪ **Make suitable arrangements for receiving and unloading the consignment at Project site**

The route inside the site and place for unloading the consignment needs to be clearly planned and cleared. It is essential that after unloading, there is no or minimum shifting required before erection. The trailer route should be such that the load is brought to unloading point and without difficulty the trailer moves out after unloading.

5.2   Logistics- Challenges

There are many challenges Logistics Manager faces. These are compounded by the conditions in developing country like India. With different monsoon periods across country, the weather conditions need to be taken into account while taking a decision.

> ▪ **Port, Road and Rail Infrastructure inadequacies**

Ports are divided into major port and minor ports. Big ships cannot berth in minor ports while major ports are congested. Barges are required for unloading consignments from ships at minor ports and bringing them to shore. At minor ports availability of suitable cranes also is an issue.

Bridges on railway and roads are very old and unsafe for heavy consignments. Width of roads is a major constraint in addition to the condition of roads.

Lack of Rail Over bridges is also a big factor in movement of consignments.

> ▪ **Permit system and Entry Tax**

Vehicles need permit across the country and these are costly. Lot of time is wasted at every state border crossing showing the permit and getting temporary permit.

Many states have entry tax and these documents need to accompany the material; otherwise the consignment is held up at state border.

> ▪ **Multiple approvals to be obtained**

For crossing every state, separate approval is required. In addition to this approval from railways is required for crossing railway lines in electrified sections.

> - **Lack of quality transport equipment**

Many trailers are fabricated by road side fabricators not giving enough attention to safety requirements. There is no formal awareness training to drivers transporting the material about cost impact of their negligence on a project.

5.3   Descriptive Questions

1. Describe the various processes involved in Logistics Management

2. Describe the steps in Transport route finalisation upto order placement for Over Dimensional Consignment (ODC)

3. What are the challenges faced in Logistics for a project?

5.4   Multiple Choice Questions

State True or False

SL	Statement	True	False
01	Logistics involves supply from all vendor to the project		
02	Route survey is carried out for major equipment		
03	Underwriters are not involved in Logistics		
04	Railway is a preferred mode of transport		

CONSTRUCTION EQUIPMENT

Construction equipment is the equipment used during construction. The details of equipment are given with individual construction activity. Here we will discuss how these are classified and basis of their selection.

Selection of construction equipment is based on

6.1 Product Design

This plays an important role in selection of construction equipment. The erection sequence for a boiler with top supported pressure parts is different than a boiler with bottom supported pressure parts. Consequently the method and type of construction will change.

The precautions required and methods for inserting a generator rotor into stator are different for horizontal axis and vertical axis generators.

6.2 Plant Layout

Plant layout defines the access for placing the equipment in position. Depending on the access availability method of erection/ construction and consequently equipment selection will change.

- Customer Facilities

    If construction power is not arranged by customer, then DG sets would be required with higher capacity or more in numbers. Instead of electrically operated equipment, diesel engine driven equipment would be required.

    The capacity of EOT crane in TG building would define the method of erection of TG set.

6.3   Material to be handled – type, weight, size

This would define the capacity of equipment and type of accessories required for safe handling of the material.

■ Output requirement

Project schedule defines the rate of erection/ construction per day or per month. Considering the efficiency and periods of outage due to maintenance or breakdown, equipment needs to be selected. If the concreting output of 3000 Cubic Metre per month is required, a batching plant of 30 Cubic Metre per hour is ideally sufficient. But in reality it would not meet the output requirements, as there are days when front is not ready and on other days the requirement would be to pour more than 300 Cubic Metre concrete. On some days when mass concreting of more than 1000 Cubic Metre has to be done in short span this capacity will be grossly inadequate. Considering this, it would be ideal to have 2 numbers 50 Cubic Metre per hour capacity batching plants.

Similarly other equipment requirement is also not a constant and likely to vary, so peak output requirement needs to be considered.

6.4   Ready availability

Desired equipment should be available when required as per project schedule. Own equipment's are sometimes preoccupied at different project site and not available. The desired rating equipment may not be readily available on required dates on hire. In such case selection of equipment has to be based on ready availability in market both of main equipment and spare parts.

6.5   Redundancy requirement

Equipment associated with critical activities need to be available on demand. To ensure this, additional number need to be planned than exact requirement to build up redundancy.

6.6   Financial Implication

This is most important aspect in equipment selection. If the project cash inflow cannot support acquisition of some costly equipment for the project, then alternative solutions have to be found out. For heavy transformer erection, placing a high capacity crane becomes uneconomical if the crane is otherwise not available at the project site. In such case conventional method of using jacks, sleepers and rollers need to be considered.

There are many ways in which equipment's are categorized. Categorisation helps in computerization of the assets and easy report generation. The equipment is categorized as follows

- **By Size or Capacity – Heavy, Medium etc**
- **By use in Mechanical erection, Electrical work, Controls & Instrumentation works and Civil works**
- **By Activity- Lifting, Earthwork, Transport etc**
- **By type of Drive- Electrical, Pneumatic, Hand operated etc.**
- **Stationery, Mobile and Portable**

The other method of categorisation is by type as

6.7   Cranes

- ✓ Gantry Cranes
- ✓ EOT Cranes

These are further classified as pendant operated or cabin operated.

- ✓ Portal Cranes
- ✓ Mobile Cranes – Crawler or Tyre mounted

These are also further classified as Lattice boom or hydraulic boom cranes.

- ✓ Tower Cranes
- ✓ Jib cranes

6.8   Lifting Tools and Tackles

- ✓ Winches
- ✓ Strand Jack equipment
- ✓ Pulley Blocks & Chain Pulley blocks
- ✓ Pullers
- ✓ Jacks – Hydraulic and screw
- ✓ Rollers
- ✓ Hammer, Spanner sets

6.9   Measuring Equipment

Selection of measuring equipment decides on the measurement taken and tolerance specified in drawing. Accordingly precision level has to be selected.

- ✓ Steel tapes – of various lengths upto 30 Metres
- ✓ Theodolite
- ✓ Total Station
- ✓ Testing equipment for gauges
- ✓ Megars – Hand operated and motorised
- ✓ Multi meters – both AC and DC
- ✓ Gauges
- ✓ Vibration measuring instruments
- ✓ Vernier caliper
- ✓ Micro meters
- ✓ Spirit Levels – to various accuracy levels

✓ Surface Table – These are required of various sizes to check matching of surfaces.

✓ Feeler gauges – of various thicknesses and lengths

✓ Dumpy level

✓ Thickness measurement gauge for paints

✓ Hardness testers – Both Lab model and portable models for weld joint checking.

6.10 Material Movement

✓ Trucks

✓ Tempos

✓ Open jeeps / MUVs

✓ Tractors

✓ Hydras – Though this is a crane, it is used for transporting material over a short distance.

✓ Dumpers – of various capacities

✓ Trailers – Low Bed, Long Bed etc

✓ Fork Lifts – are rarely used in Indian sites

✓ Horse – It is the driver cabin only which can be attached to a trailer. This way the horse can take a trailer to a destination and leave it to get unloaded and take another trailer for movement.

✓ Loader

- **Concrete Mixers, Batching Plant**
- **Excavators**
- **Service equipment**

✓ Air compressors – Trolley mounted air compressors are used to clean parts before erection and to ensure that tubes are not blocked. In civil works, breakers use air compressors.

✓ Water Pumps – These can be reciprocating or centrifugal pumps. Reciprocating pumps are used for hydraulic tests of pressure parts. Both electrically operated and diesel engine driven pumps are used.

✓ Hand operated pumps – mainly for emptying barrels.

✓ DG sets – Both trolley mounted DG sets and skid mounted DG sets of various capacities used.

✓ Fire Tender

✓ Chilling/ Ice plant – It is required where temperature controlled concreting is involved.

6.11   Metal forming & Cutting Tools

✓ Pipe Bending machines – Both hand operated and hydraulic bending machines are used for site routed piping.

✓ Grinders – Pedestal & Portable

✓ Gas cutting torches

✓ Sheet cutting & forming tools

✓ Hammers

✓ Chisels

✓ Drilling Machines – Portable machines used for small diameter holes. Machines with magnetic base are used for in situ drilling. Pedestal machines used in workshop.

✓ Lathes

✓ Milling Machines

✓ Saws

✓ Bending machines for rebars

6.12   Electrical equipment

✓ Termination tools

✓ Megars – Both hand operated and motorized Megars are used.

✓ Multi meters

✓ Hi Pot testing kit

✓ Phase sequence meter

✓ Tong tester

✓ Termination kits

6.13   Laboratory equipment

Laboratory equipment

✓ Universal Testing Machine

✓ Concrete cube moulds

✓ Curing tanks

✓ Bending machine

✓ Cones for slump test

✓ Weighing machine

6.14   Welding machines

✓ Transformer or Generator

✓ Gas Welding

✓ TIG/ MIG/ CO2 welding

✓ Soldering and Brazing equipment

6.15   Heat Treatment & Radiography equipment

✓ Developing & viewing equipments for radiograph films

✓ Heating elements and temperature gauges & recorders

✓ Stress relieving equipment for special material – Material like T91/ T92 have special requirements and cannot be stress relieved by heating elements.

6.16   Descriptive Questions

1. Describe the process of Equipment identification and deployment for a project

2. What is History card? Explain in short the contents of History card and its importance.

3. What is NABL? Describe the importance of calibration.

4. When do you declare an equipment as BER?

6.17   Multiple Choice Questions

State True or False

SL	Statement	True	False
01	You receive a consignment of new chain pulley block. You need to get it inspected by a Competent person before use		
02	You receive a new Hydra crane at site without registration number. It can be used immediately		
03	Customer facilities has no effect on equipment selection		
04	Plant layout defines selection of equipment		
05	Product design has no impact on equipment selection		
06	For major equipment Buy or Hire exercise needs to be carried out		
07	Equipment History card has no use		

NON DESTRUCTIVE TESTING/ EXAMINATION

Nondestructive testing (NDT) is the process of inspecting, testing, or evaluating materials, components or assemblies for discontinuities, or differences in characteristics without destroying the serviceability of the part or system. In other words, when the inspection or test is completed the part can still be used.

In contrast to NDT, other tests are destructive in nature and are therefore done on a limited number of samples ("lot sampling"), rather than on the materials, components or assemblies actually being put into service.

These destructive tests are often used to determine the physical properties of materials such as impact resistance, ductility, yield and ultimate tensile strength, fracture toughness and fatigue strength, but discontinuities and differences in material characteristics are more effectively found by NDT.

7.1   NDT Test methods

Test method names often refer to the type of penetrating medium or the equipment used to perform that test. Current NDT methods are:

✓ Acoustic Emission Testing (AE)

It is a method that is used to analyse emitted sound waves caused by defects or discontinuities. These acoustic waves are induced by small deformations, corrosion or cracking, which occur prior to structure failure. It is therefore possible, with AET, to locate structural defects and to monitor the propagation and development of discontinuities.

AET is a method which evaluates the elasticity of waves caused by discontinuities formed within the specimen. In large-sized structures, several sensors are placed on the material surface, leaving a space of some metres in between. The information collected by each of the sensors is monitored through a computer.

✓ Electromagnetic Testing (ET)

It is the process of inducing electric currents or magnetic fields or both inside a test object and observing the electromagnetic response. If the test is set up properly, a defect inside the test object creates a measurable response. It is used for the detection of surface connected flaws. The results are qualitative and based upon similar responses to known reflectors in calibration standards. This NDT technique is ideal for both ferrous and non-ferrous parts and is capable of inspection through coatings that are intact and smooth.

The term "Electromagnetic Testing" is often intended to mean simply Eddy-Current Testing (ECT). However with an expanding number of electromagnetic and magnetic test methods, "Electromagnetic Testing" is more often used to mean the whole class of electromagnetic test methods, of which Eddy-Current Testing is just one.

✓ Laser Testing Methods (LM)

Laser ultrasonic testing (LUT) is a remote, noncontact extension of conventional, contact or near-contact ultrasonic testing (UT). A laser pulse is directed to the surface of a sample through a fibre or through free space. The laser pulse interacts at the surface to induce an ultrasonic pulse that propagates into the sample. This ultrasonic pulse interrogates a feature of interest and then returns to the surface. A separate laser receiver detects the small displacement that is generated when the pulse reaches the surface. The electronic signal from the receiver is then processed to provide the measurement of interest. Laser UT is fast and effective on rough surfaces. It is ideally suited for many applications that are beyond the capabilities of conventional ultrasonic testing.

It is used for

- Wall thickness measurement

- Weld inspection

- Coating thickness measurement

- Composite flaw detection

- Crack depth measurement

- Bond evaluation

- Grain size measurement

✓ Leak Testing (LT)

Leak detection includes hydrostatic test after erection and leak detection during service.

✓ Magnetic Flux Leakage (MFL)

It is used to detect corrosion and pitting in steel structures, most commonly pipelines and storage tanks. The basic principle is that a powerful magnet is used to magnetize the steel. At areas where there is corrosion or missing metal, the magnetic field "leaks" from the steel. In an MFL tool, a magnetic detector is placed between the poles of the magnet to detect the leakage field. Analysts interpret the chart recording of the leakage field to identify damaged areas and hopefully to estimate the depth of metal loss.

✓ Liquid Penetrant Testing (PT)

Dye penetrant inspection (DPI), also called liquid penetrant inspection (LPI) or penetrant testing (PT), is a widely applied and low-cost inspection method used to locate surface-breaking defects in all non-porous materials (metals, plastics, or ceramics). The penetrant may be applied to all non-ferrous materials and ferrous materials; although for ferrous components magnetic-particle inspection is often used instead for its subsurface detection capability. LPI is used to detect casting, forging and welding surface defects such as hairline cracks, surface porosity, leaks in new products, and fatigue cracks on in-service components.

DPI is based upon capillary action, where a low surface tension fluid penetrates into clean and dry surface-breaking discontinuities. Penetrant may be applied to the test component by dipping, spraying, or brushing. After adequate penetration time has been allowed, the excess penetrant is removed and a developer is applied. The developer helps to draw penetrant out of the flaw so that an invisible indication becomes visible to the inspector. Inspection is performed under ultraviolet or white light, depending on the type of dye used - fluorescent or nonfluorescent (visible).

✓   Magnetic Particle Testing (MT)

Magnetic particle Inspection (MPI) is a non-destructive testing (NDT) process for detecting surface and slightly subsurface discontinuities in ferromagnetic materials such as iron, nickel, cobalt, and some of their alloys. The process puts a magnetic field into the part. The piece can be magnetized by direct or indirect magnetization. Direct magnetization occurs when the electric current is passed through the test object and a magnetic field is formed in the material. Indirect magnetization occurs when no electric current is passed through the test object, but a magnetic field is applied from an outside source. The magnetic lines of force are perpendicular to the direction of the electric current which may be either alternating current (AC) or some form of direct current (DC) (rectified AC).

The presence of a surface or subsurface discontinuity in the material allows the magnetic flux to leak, since air cannot support as much magnetic field per unit volume as metals. Ferrous iron particles are then applied to the part. The particles may be dry or in a wet suspension. If an area of flux leakage is present, the particles will be attracted to this area. The particles will build up at the area of leakage and form what is known as an indication. The indication can then be evaluated to determine what it is, what may have caused it, and what action should be taken, if any.

✓   Radiographic Testing (RT)

It is a non-destructive testing of components and assemblies that is based on differential absorption of penetrating radiation- either electromagnetic radiation of very short wave-lengths or particulate radiation by the part or test piece being tested. Because of differences in density and variations in thickness of the part, or differences in absorption characteristics caused by variation in composition, different portions of a test piece absorb different amounts of penetrating radiation. Unabsorbed radiation passing through the part can be recorded on film or photosensitive paper, viewed on a fluorescent screen or monitored by various types of electronic radiation detectors.

The term radiography testing usually implies a radiographic process that produces a permanent image on film or paper.

Before commencing a radiographic examination, it is always advisable to examine the component with one's own eyes, to eliminate any possible external defects. If the surface of a weld is too irregular, it may be desirable to grind it to obtain a smooth finish.

The beam of radiation must be directed to the middle of the section under examination and must be normal to the material surface at that point, except in special techniques where known defects are best revealed by a different alignment of the beam. The length of weld under examination for each exposure shall be such that the thickness of the material at the diagnostic extremities, measured in the direction of the incident beam, does not exceed the actual thickness at that point by more than 6%. The specimen to be inspected is placed between the source of radiation and the detecting device, usually the film in a light tight holder or cassette, and the radiation is allowed to penetrate the part for the required length of time to be adequately recorded.

The result is a two-dimensional projection of the part onto the film, producing a latent image of varying densities according to the amount of radiation reaching each area. It is known as a radio graph, as distinct from a photograph produced by light. Because film is cumulative in its response (the exposure increasing as it absorbs more radiation), relatively weak radiation can be detected by prolonging the exposure until the film can record an image that will be visible after development. The radiograph is examined as a negative, without printing as a positive as in photography. This is because, in printing, some of the detail is always lost and no useful purpose is served.

Safety is of prime importance and adequate precautions need to be taken. In India the radiography source and radiography is controlled as per Radiation Protection Rules 2004. Normally radiography is carried out at construction sites in night when workers are not around to reduce chance of exposure to radiation.

✓ Thermal/Infrared Testing (IR)

Thermographic inspection refers to the nondestructive testing of parts, materials or systems through the imaging of the thermal patterns at the object's surface. **Infrared thermography** is a non-destructive, non-intrusive, non-contact mapping of thermal patterns or "thermograms", on the surface of objects. An infrared thermographic scanning system can measure and view temperature patterns based upon temperature differences as small as a few hundredths of a degree Celsius. Infrared thermographic testing may be performed during day or night, depending on environmental conditions and the desired results.

A typical application for IR Thermographic equipment is looking for "hot spots" in electrical equipment, which illustrates high resistance areas in electrical circuits.

✓ Ultrasonic Testing (UT)

It is a family of non-destructive testing techniques based on the propagation of ultrasonic waves in the object or material tested. In most common UT applications, very short ultrasonic pulse-waves with center frequencies ranging from 0.1-15 MHz, and occasionally up to 50 MHz, are transmitted into materials to detect internal flaws or to characterize materials. A common example is ultrasonic thickness measurement, which tests the thickness of the test object, for example, to monitor pipework corrosion.

Ultrasonic testing is often performed on steel and other metals and alloys, though it can also be used on concrete, wood and composites, albeit with less resolution.

In ultrasonic testing, an ultrasound transducer connected to a diagnostic machine is passed over the object being inspected. The transducer is typically separated from the test object by a couplant (such as oil) or by water, as in immersion testing. However, when ultrasonic testing is conducted with an Electromagnetic Acoustic Transducer (EMAT) the use of couplant is not required.

There are two methods of receiving the ultrasound waveform: reflection and attenuation. In reflection (or pulse-echo) mode, the transducer performs both the sending and the receiving of the pulsed waves as the "sound" is reflected back to the device. Reflected ultrasound comes from an interface, such as the back wall of the object or from an imperfection within the object. The diagnostic machine displays these results in the form of a signal with an amplitude representing the intensity of the reflection and the distance, representing the arrival time of the reflection. In attenuation (or through-transmission) mode, a transmitter sends ultrasound through one surface, and a separate receiver detects the amount that has reached it on another surface after travelling through the medium. Imperfections or other conditions in the space between the transmitter and receiver reduce the amount of sound transmitted, thus revealing their presence.

Using the couplant increases the efficiency of the process by reducing the losses in the ultrasonic wave energy due to separation between the surfaces.

Rolled steel products are inspected with UT before accepting them for fabrication.

✓ Vibration Analysis (VA)

Vibration Analysis is used to detect early precursors to machine failure, allowing machinery to be repaired or replaced before an expensive failure occurs.

It is known that rotating equipments vibrate to some extent. Similarly every structure and foundation has a natural frequency. When the rotating equipment vibration matches natural frequency of foundation there is resonance which is harmful to both equipment and foundation. By balancing the rotating equipment the vibrations are brought down to acceptable limit.

✓ Visual Testing (VT)

A visual inspection or visual examination of objects, parts or components is the oldest and reliable non-destructive testing method. The test method is applied to almost every product as a quality assurance tool. It is also cheapest test method.

✓ Rebound Hammer test for concrete

In case the concrete cubes fail, concrete test hammer may be used to arrive at strength of the concrete laid. The hammer consists of a spring controlled mass that slides on a plunger within a tubular housing. When the plunger is pressed against, the surface of concrete, it retracts against the force of the spring. When completely retracted the spring is automatically released. On the spring controlled mass rebound, it takes the rider with it along the guide scale. By pushing a button, the rider can be held in position to allow readings to be taken. Each hammer varies considerably in performance and needs calibration for use on concrete made with aggregates produced from a specific source. For details refer relevant Indian Standard and ASTM C805.

7.2 Personnel Qualification

As specialized skills are required to conduct the tests and interpret them, companies conduct in-house training programs. Training programs and Standards by American Society for Nondestructive Testing (ASNT) are used to qualify the personnel and accepted in the industry and regulators.

7.3   Levels of certification

Most NDT personnel certification schemes listed above specify three "levels" of qualification and/or certification, usually designated as *Level 1*, *Level 2* and *Level 3* (although some codes specify Roman numerals, like *Level II*). The roles and responsibilities of personnel in each level are generally as follows (there are slight differences or variations between different codes and standards):

- **Level 1** are technicians qualified to perform only specific calibrations and tests under close supervision and direction by higher level personnel. They can only report test results. Normally they work following specific work instructions for testing procedures and rejection criteria.

- **Level 2** are engineers or experienced technicians who are able to set up and calibrate testing equipment, conduct the inspection according to codes and standards (instead of following work instructions) and compile work instructions for Level 1 technicians. They are also authorized to report, interpret, evaluate and document testing results. They can also supervise and train Level 1 technicians. In addition to testing methods, they must be familiar with applicable codes and standards and have some knowledge of the manufacture and service of tested products.

- **Level 3** are usually specialized engineers or very experienced technicians. They can establish NDT techniques and procedures and interpret codes and standards. They also direct NDT laboratories and have central role in personnel certification. They are expected to have wider knowledge covering materials, fabrication and product technology.

7.4   Descriptive Questions

1. Describe the Non Destructive Tests/ Examinations involved for a project.

2. What is ASNT? Explain the different levels of certification by ASNT and their importance in working.

7.5   Multiple Choice Questions

State True or False

SL	Statement	True	False
01	Anyone can review and accept a radiograph		
02	Hammer rebound test is used for concrete		
03	LPI is used to detect leaks in a system		
04	Ultrasonic test is used in structure testing		

PROJECT MANAGEMENT BASICS

Though the Project level schedule is prepared by Customer/ Consultant or EPC contractor, Construction Company needs to prepare schedules and networks for their scope of work. For EPC contractor or Construction contractor, awareness of basics of Project Management techniques is essential.

Project Management is the Discipline of planning, organizing, securing, managing, leading, and controlling resources to achieve specific goals.

It is a control process to verify various processes in a Project for

 ➢ Adherence to Schedule

 ➢ Deployment and Utilisation of Resources

 ➢ Deliverables

 ➢ Cost

The need for Project Management has arisen due to

 ➢ Long Cycle Times – More than a year for all projects and in some cases many years required for completion.

 ➢ High Investment – Thousands of Crores are involved in modern projects

➤ Complexities – With technology development in all the fields and stricter norms by regulators, the complexities have multiplied. There are many interdependent activities in a project and the ownership is with different agencies. One agency's progress affects work of other agencies.

➤ Involvement of multiple agencies – The number of agencies involved is large in a big project and it is difficult to check individually the effect on other agencies.

➤ High Risks – Risks are both financial and regulatory of nature. With higher investment the risk potential is also high.

Project Management was recognised as a management tool in 1950s. Till 1950s Gantt charts and other tools were used for monitoring projects. Critical Path Method (CPM) and Project Evaluation and Review Technique (PERT) were developed in 1950s.

ISO 10006 gives guidance on application of quality management techniques in projects. ISO 21500 has been published as **"Project Management - Guide to project Management".**

PERT chart pictorially depicts the relationship between activities

It gives activity duration, Early Start, Early Finish; Late start, Late Finish

---

- **Two methods in vogue for drawing a PERT chart**

➤ Activity on line – In this method the activity number would be a combination of two node numbers it is connecting typically i(n),j(n).

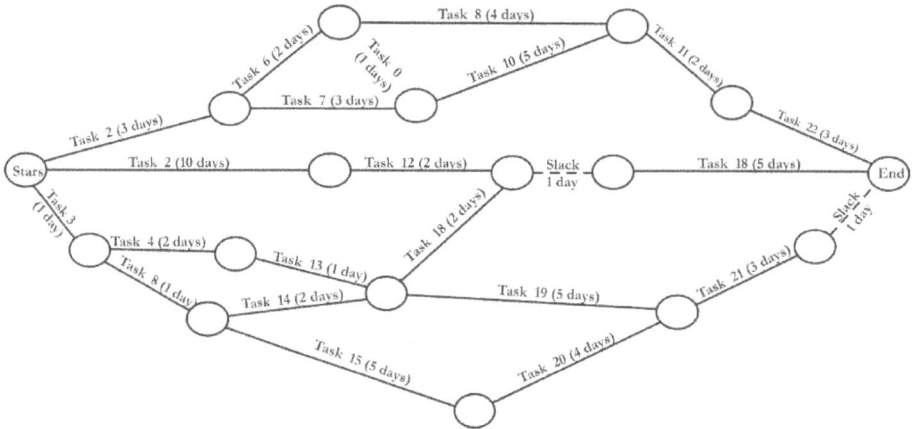

> Activity on node – In this method each activity is represented by a single number.

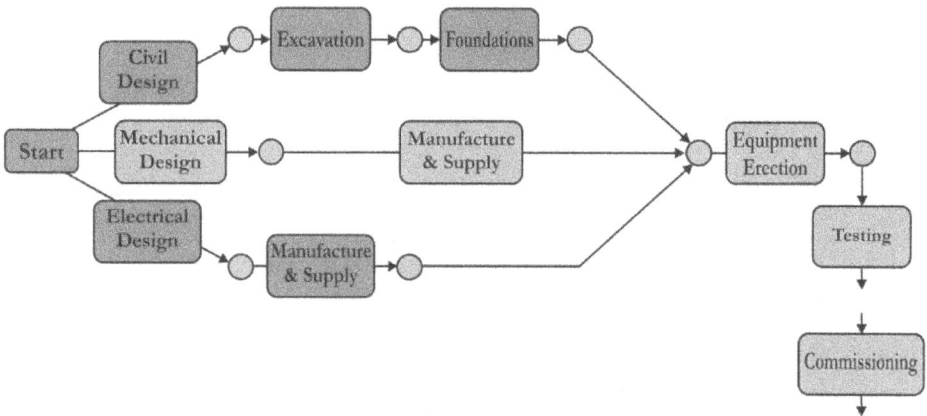

Most of commercial softwares use Activity on node method for drawing PERT chart. For fixing the dates of Start/ Finish of activities, Forward and Backward calculations are done with activity durations as input. The relation between activities decides their start and completion dates.

The terms used in PERT chart are

- **Early Start – When an Activity can start at the earliest. This is calculated in the forward pass, starting from Project start date.**

- Early Finish – When an Activity is expected to complete at the earliest. This is calculated in the forward pass, starting from Project start date.

- Duration – It is Time required to complete an Activity in days or weeks.

- Late Start - When an Activity has to start at the latest without affecting schedule. This is calculated in the backward pass, starting from Project completion date.

- Late Finish - When an Activity has to finish at the latest without affecting schedule. This is calculated in the backward pass, starting from Project completion date.

- Float – Time between Early Finish and Late Finish

- Dummy Activity – An Activity with zero duration. This concept is used in PERT charts where activity is on line.

- Activity code – An Alphanumeric number to identify an Activity. This helps in segregating activities.

- Activity Description – Based on software used, the number of characters used to describe the activity change.

- Critical Activity – An Activity with zero float.

- Critical Path – A Path that connects Critical Activities. It is the longest path in a Network

- Milestone Activity – Important activities and dates for a project. As a practice Milestone activities start and complete on the same date.

- Event – Other than Milestone activities, these are important activities of a project which are monitored for a package/ island.

- Constraint – External input to ensure date conformance

- Owner Input – Activities outside network.

- Actual Start – User input

- Actual Completion - User input

- % Completion - User input

- Work Break down Structure (WBS) – It is used to segregate activities for easy monitoring.

- **Lag – Time difference between start of two activities, necessitated by logic. In a long cycle activity involving many items, this helps in reducing the overall duration of project. After welding certain number of joints, radiography and stress relieving can be started without waiting for completing all welding.**

- **Predecessor – Activity that precedes current activity**

- **Successor – Activity that succeeds current activity**

In a Project there is only one activity which has no predecessor i.e. Start of Project. Similarly there is only one activity that does not have successor and that is Project Completion. It has to be ensured that this condition is achieved. Similarly while connecting the activities; it has to be ensured that a successor is not a predecessor to an earlier activity. This is called a loop and while checking logic of network this needs to be ensured.

Activities are connected to each other in different ways. The way activities are connected to each other defines the project time. It is called the logic of network.

➢ Finish to Start – Activities are in series. This means only after completing earlier activity next activity can start.

➢ Start to Start – Activities are in parallel. Activities are independent of each other and can start simultaneously.

➢ Finish to Finish – An Activity cannot be completed till other is completed. Consider the earlier example of welding; Radiography and stress relieving activities cannot be finished before welding is completed.

- **Project Calendar – This defines the week numbers for the project. In the Project calendar weekly offs, number of hours per shift, number of shifts in a day and public holidays are defined. It has impact on project completion and utmost care needs to be taken.**

8.1   Work Breakdown Structure

A **work breakdown structure (WBS)**, in Project Management and Systems Engineering, is a deliverable oriented decomposition of a Project into smaller components. It defines and groups a Project's discrete work elements in a way that helps organize and define the total work scope of the Project.

A work breakdown structure element may be a product, data, service, or any combination. A WBS also provides the necessary framework for detailed cost estimating and control along with providing guidance for schedule development and control.

WBS is a hierarchical decomposition of the project into phases, deliverables and work packages. It is a tree structure, which shows a subdivision of effort required to achieve an objective; for example a program, project, and contract. In a project or contract, the WBS is developed by starting with the end objective and successively subdividing it into manageable components in terms of size, duration, and responsibility (e.g., systems, subsystems, components, tasks, subtasks, and work packages) which include all steps necessary to achieve the objective.

A work breakdown structure permits summing of subordinate costs for tasks, materials, etc., into their successively higher level "parent" tasks, materials, etc. For each element of the work breakdown structure, a description of the task to be performed is generated. This technique (sometimes called a system breakdown structure) is used to define and organize the total scope of a Project.

For instance, a Boiler is divided into Structure, Pressure Parts, Non Pressure Parts and Rotating Equipment etc. Each of these can be subdivided into sub assemblies and attach activities like engineering, procurement, manufacturing etc to them.

The WBS is not an exhaustive list of work. It is instead a comprehensive classification of project scope. WBS is neither a project plan; a schedule, nor a chronological listing. It specifies what will be done, not how or when.

There are certain rules that need to be followed while preparing a WBS.

> An important design principle for work breakdown structures is called the 100% rule. It has been defined as follows:

The 100% rule states that the WBS includes 100% of the work defined by the project scope and captures all deliverables – internal, external, interim – in terms of the work to be completed, including project management. The 100% rule is one of the most important principles guiding the development, decomposition and evaluation of the WBS. The rule applies at all levels within the hierarchy: the sum of the work at the "child" level must equal 100% of the work represented by the "parent" and the WBS should not include any work that falls outside the actual scope of the project, that is, it cannot include more than 100% of the work. It is important to remember that the 100% rule also applies to the activity level. The work represented by the activities in each work package must add up to 100% of the work necessary to complete the work package.

> ➢ Mutually exclusive: In addition to the 100% rule, it is important that there is no overlap in scope definition between two elements of a work breakdown structure. This ambiguity could result in duplicated work or miscommunications about responsibility and authority. Such overlap could also cause confusion regarding project cost accounting. If the WBS element names are ambiguous, a WBS dictionary can help clarify the distinctions between WBS elements. The WBS Dictionary describes each component of the WBS with milestones, deliverables, activities, scope, and sometimes dates, resources, costs, quality.

➢ **Plan outcomes, not actions**

- If the work breakdown structure designer attempts to capture any action-oriented details in the WBS, s/he will likely include either too many actions or too few actions. Too many actions will exceed 100% of the parent's scope and too few will fall short of 100% of the parent's scope.

- The best way to adhere to the 100% rule is to define WBS elements in terms of outcomes or results, not actions. This also ensures that the WBS is not overly prescriptive of methods, allowing for greater ingenuity and creative thinking on the part of the project participants. Work breakdown structures that subdivide work by project phases (e.g. preliminary design phase, critical design phase) must ensure that phases are clearly separated by a deliverable also used in defining entry and exit criteria (e.g. an approved preliminary or critical design review).

➢ **Level of detail**

- One must decide when to stop dividing work into smaller elements. This will assist in determining the duration of activities necessary to produce a deliverable defined by the WBS. There are several heuristics or "rules of thumb" used when determining the appropriate duration of an activity or group of activities necessary to produce a specific deliverable defined by the WBS.

- The first is the "80 hour rule" which means that no single activity or group of activities to produce a single deliverable should be more than 80 hours of effort.

- The second rule of thumb is that no activity or series of activities should be longer than a single reporting period. Thus if the project team is reporting progress monthly, then no single activity or series of activities should be longer than one month long.

- The last heuristic is the "if it makes sense" rule. Applying this rule of thumb, one can apply "common sense" when creating the duration of a single activity or group of activities necessary to produce a deliverable defined by the WBS.

- A work package at the activity level is a task that:

✓ can be realistically and confidently estimated;

✓ makes no sense practically to break down any further;

✓ can be completed in accordance with one of the heuristics defined above;

✓ produces a deliverable which is measurable; and

✓ Forms a unique package of work which can be outsourced or contracted out.

➢ **Coding scheme**
- It is common for work breakdown structure elements to be numbered sequentially to reveal the hierarchical structure. The purpose for the numbering is to provide a consistent approach to identifying and managing the WBS across like systems regardless of vendor or service. For example 1.1.2 ID Fan identifies this item as a Level 3 WBS element, since there are three numbers separated by a decimal point. A coding scheme also helps WBS elements to be recognized in any written context.

➢ **Terminal element**
- The lowest elements in a tree structure, a terminal element is one that is not further subdivided.

- In a Work Breakdown Structure such (activity or deliverable) elements are the items that are estimated in terms of resource requirements, budget and duration; linked by dependencies; and scheduled. At the juncture of the WBS element and organization unit, control accounts and work packages are established and performance is planned, measured, recorded and controlled. A WBS can be expressed down to any level of interest.

- Three levels are the minimum recommended, with additional levels for and only for items of high cost or high risk, and two levels of detail at cases such as systems engineering or program management, with the standard showing examples of WBS with varying depth such as software development at points going to 5 levels or fire-control system to 7 levels.

➤ **Consistent to Norms**

The higher WBS structure should be consistent to whatever norms or template exists within the organization or domain. In other words, it should be consistent with the practice in the particular industry.

8.2   Activity weightages

It is important to know the project progress at any instance. To facilitate this weightages are assigned to WBS elements and further divided into sub elements. Depending on Scope of Work, Packages and Generic activities are assigned weightages. These are further divided into sub activities. For instance Engineering may be assigned a weightage of 20% for the entire project. Considering Engineering as 100%, it is further divided into Boiler, TG and Electrical and so on. This way the overall progress can be derived very easily.

8.3   Descriptive Questions

1. Describe the various terms used in a project management network and their meaning.

2. Describe various methods of connecting activities and the impact of each connection on network.

3. What is WBS? Explain its importance in network and various rules followed in WBS

4. Why activities need to be assigned weightage. Explain the methods followed and its impact on progress report based on weightages.

8.4   Multiple Choice Questions

State True or False

SL	Statement	True	False
01	Work Breakdown Structure is important aspect in Project Management		
02	Project Calendar assigns week numbers		
03	There can be multiple activities in a network without successor		
04	A network can have multiple critical paths		
05	Plan outcomes and not actions		

CONSTRUCTION – CIVIL WORKS ACTIVITIES

Civil construction in India has to abide by Indian Standards. Standards for cement, water, sand, aggregate, steel and all other material used are available and a practicing civil engineer needs to be familiar with these standards. Work practices also need to be compliant with Indian Standards.

- Cleaning and clearing the site – This involves clearing grass, vegetation, boulders and bushes from work site and make it ready for further activities. Permission from Forest department needs to be taken before cutting trees/ bushes.

- Area grading and leveling – In earlier projects, this was carried out before giving clearance to individual contractors. However nowadays to save overall cost, this activity is not carried out. Contour of site is modified only for ease of movement and some high points dressed to merge with overall contour.

- Contour checking and putting bench marks – Levels are marked on Plot plan with respect to Mean Sea Level (MSL). The site is divided based on Longitude and Latitude. Instead of referring to these complex numbers, the site is divided in 100 x 100 meter squares with centerline of plot earmarked as zero and then naming each grid as East 100, West 100 and similarly in North/ South direction. All foundations and structures have to comply with design requirement. After initial survey, bench marks are constructed using bricks at strategic locations and their levels marked. All further works are carried out using these benchmarks and confirming levels of foundations. Each benchmark carries the level and grid marking also.

- Setting out – This is the first activity when the outline of foundation is marked on the ground as per Plot plan. Temporary benchmark for this foundation is made so that for every measurement one need not refer to permanent benchmarks. Location of this benchmark is such that it does not get disturbed during excavation or further activities.

- Excavation – Excavation means digging and removal of soil. Before one starts excavation, it is essential to confirm that there are no underground facilities like pipes, cables, trenches etc. It is also necessary to put barricades and caution signs around the excavation area so that no one falls into the pit by accident. To ensure this a work permit system is used in construction sites. Mechanized excavation and transportation of soil was introduced in 1980s in India. Prior to this period it was labour intensive activity and donkeys were used to transport excavated material.

Before commencing excavation, area for dumping excavated earth and the route to be followed needs to be identified. Part of excavated earth can be used for back filling the area around foundations, if it meets specifications. The distance between source and dumping area is called lead. Every contract specifies the lead for which no payment is made for transportation. If the distance exceeds the contract specification, additional payment needs to be made to the contractor.

Two terms need to be understood:

- NGL or Natural Ground Level – The ground level as it exists before commencing excavation, if the area is not graded.

- FGL or Finished Ground Level – The level that will be attained after completing works and area cleared for operation. Normally top of concrete floor level.

Excavations need to be carried out to a specified minimum depth from NGL.

Excavation charges depend on the type of soil excavated. Soil is classified as Black cotton soil, Murrum, Soft rock, Medium rock and Hard rock. The equipment used and rate of excavation depend on the type of soil. While estimating the time required for excavation, type of soil needs to be considered. For breaking hard rock earlier blasting was used. With restrictions on blasting, open blasting was replaced by controlled blasting. For controlled blasting also permission needs to be obtained. Permission is required for handling and storing explosives. Licensed contractors are used for carrying out blasting.

When excavations go deep, there is a chance of the side walls collapsing and people getting entrapped in under the burden. To avoid such eventuality, Bureau of Indian Standards has issued a standard that specifies slopes to be maintained in different types of soil. A ramp needs to be provided for going down and coming up for men and equipments.

Ground Water Table affects excavation activity. This is the depth at which water will be found. It depends on site location and existence of water bodies around the plant. When excavation goes beyond water table level, one can see water oozing from side walls and the excavated pit gets flooded. To avoid flooding, it is necessary to give slope to bottom of excavated area and make a drain pit large enough to collect all water and keep excavated area dry. Collected water needs to be pumped out regularly so that work can continue uninterrupted. Arrangement for draining such water needs to be made from all areas of plant.

After reaching the design depth of foundation, Soil Bearing Capacity (SBC) needs to be checked. If SBC is lower than Design requirement, excavation needs to be carried out further to attain requisite value. This will impact the design of foundation and revised design is required.

- PCC or mud mat – Plain cement concrete (PCC) is used to provide rigid impervious bed to RCC in *foundation* where the earth is soft and yielding. It is also used in other types of earth to provide a level surface under RCC so that reinforcements can be tied properly.

  Specification of PCC is as per design requirement. In case of presence of water at bottom level, additional amount of cement needs to be added to compensate for the possible loss due to dilution by flooding.

- Reinforcement preparation and placement – Concrete has good compressive strength but lacks tensile strength. Tensile strength is added by introduction of reinforcement by way of steel. Steel used can be plain rods or twisted rods.

Steel wires are used in pre stressed concrete structures in addition to reinforcement steel.

Reinforcement steel comes in standard sizes and needs to be cut to suit the design requirement. To reduce wastage, cutting plan needs to be prepared for every size of material. Before cutting the rods, it is essential to ensure that material meets the specification. For this Factory test certificates are reviewed vis-a-vis the design specification and applicable standards. In case of doubt, destructive testing needs to be done again before use.

Rods are cut as per cutting plan and then bent as per drawing. Electrically operated Bending machines are used. For smaller sizes of rods and lesser quantity, manual bending is also practiced. After preparing the reinforcement as per drawing, it is transported to work spot for placement. To maintain a uniform gap at bottom mini cubes of 50mm are placed below the reinforcement bars. During initial placement, the rods may require temporary supports to hold them in position. Binding steel wires are used to hold the bars and stirrups together. During and after assembly, the reinforcement is checked for compliance with design and standards.

▪ Inserts preparation and placement – In RCC, various types of inserts and embedment are put for use in Mechanical/ Electrical works. Curb angles are used to protect the edges of concrete structures. These inserts need to be welded with the reinforcement for keeping them firmly in position. Brackets made of reinforcement steel are first welded to the inserts and then these are secured with either binding wires or welding with reinforcement. Inserts can be plain pipes for routing cables/ piping or for inserting foundation bolts. In water tanks the inlet/ outlet pipes are put during wall construction. These are called puddle pipes. On roof of tanks there would be vents in form of inserts. These need to be put as per drawing in required orientation. To ensure this Mechanical/ Electrical Engineers are required to check the correctness with their erection drawings the placement of inserts. Normally Mechanical equipment supplier would supply these items to civil agency for incorporating in concrete.

For water retaining structures, fill bars are used at specified locations in drawing to arrest water ingress/ egress. These are made of rubber in typical shapes. These need to be put in position before taking up Form work.

- Form work preparation and placement – Form work can be made of wood/ plywood, Fiberglass or Steel plates. Corrugated sheets are used as bottom support for floor construction. Form work is also called shuttering. These sheets remain there permanently. Surface finish requirement decides the type of Form work used. Plywood can be used for maximum six operations, whereas a steel sheet can be used till it gets deformed or damaged. For plywood wooden batons are required, whereas steel plates can be bolted together using the slotted angle provided on them. Fiberglass formwork is used where modular design is used and is a latest development. It is costly but gives better surface finish and improves productivity. To fix formwork, scaffolding needs to be erected. Where height of work is more than 1.5M safe working platforms need to be erected for the workers' safety. Formwork needs to be kept in proper position by giving adequate supports. Care has to be taken to ensure that there is no bulging due to weight of concrete. This defines the number of supports and distance between them. During placement of formwork and after formwork is placed, it is inspected to meet the safety requirements in addition to design requirements. For complex structures, the design calculations for scaffolding need to be submitted for review.

- Pouring of Concrete – It involves preparing the concrete mixture, Transporting the concrete to work spot and then pouring it properly in the work area.

9.1   Nominal Mixes

In the past, the specifications for concrete prescribed the proportions of cement, fine and coarse aggregates. These mixes of fixed cement-aggregate ratio which ensures adequate strength are termed nominal mixes. These offer simplicity and under normal circumstances, have a margin of strength above that specified. However, due to the variability of mix ingredients the nominal concrete for a given workability varies widely in strength.

9.2   Standard Mixes

The nominal mixes of fixed cement-aggregate ratio (by volume) vary widely in strength and may result in under- or over-rich mixes. For this reason, the minimum compressive strength has been included in many specifications. These mixes are termed standard mixes.

IS 456-2000 has designated the concrete mixes into a number of grades as M10, M15, M20, M25, M30, M35 and M40. In this designation the letter M refers to the mix and the number to the specified 28 day cube strength of mix in N/mm2.

9.3   Designed Mixes

In these mixes the performance of the concrete is specified by the designer but the mix proportions are determined by the producer of concrete, except that the minimum cement content can be laid down. This is most rational approach to the selection of mix proportions with specific materials in mind, possessing more or less unique characteristics. The approach results in the production of concrete with the appropriate properties most economically.

In Project sites, based on design requirements Designed mixes are developed by contracting local accredited civil laboratories. These mixes take into account the quality of locally available sand and aggregate and mineral and chemical admixtures and probable cement suppliers to the project. For every cement brand, a designed mix needs to be developed for every grade of concrete. Based on test results, some cement brands can be prohibited for use at site.

Concrete can either be made at site to requirement or can be procured as ready mix concrete from an agency. The choice depends on the location of site and space availability. The choice also depends on availability of quality suppliers in the vicinity of the project and time constraints.

A **concrete mixer** (also commonly called a **cement mixer**) is a device that homogeneously combines cement, aggregate such as sand or gravel, and water to form concrete. A typical concrete mixer uses a revolving drum to mix the components. For smaller volume works, portable concrete mixers are often used so that the concrete can be made at the construction site, giving the workers ample time to use the concrete before it hardens.

Today's market increasingly requires consistent homogeneity and short mixing times for the industrial production of ready-mix concrete, and more so for precast/prestressed concrete. This has resulted in refinement of mixing technologies for concrete production. Different styles of stationary mixers have been developed, each with its own inherent strengths targeting different parts of the concrete production market. The most common mixers used today fall into 3 categories:

- Twin-shaft mixers, known for their high intensity mixing and short mixing times. These mixers are typically used for high strength concrete, RCC and SCC, typically in batches of 2–6 m³.

- Vertical axis mixers, most commonly used for precast and prestressed concrete. This style of mixer cleans well between batches, and is favoured for coloured concrete, smaller batches (typically 0.75–3 m³), and multiple discharge points. Within this category, the Pan mixers are losing popularity to the more efficient Planetary (or counter-current) mixers as the additional mixing action helps in production of more critical concrete mixes (colour consistency, SCC, etc.).

- Drum mixers (reversing drum mixer and tilting drum mixers), used where large volumes (batch sizes of 3–9 m³) are being produced. This type of mixer dominates the ready-mixed market as it is capable of high production speeds, ideal for slump concrete, and where overall cost of production is important. Drum mixers have the lowest maintenance and operating cost of the three styles of mixers.

All the mixer styles have their own inherent strengths and weaknesses, and all three styles of mixers are used throughout the world to varying degrees of popularity.

9.4   Concrete mixing transport trucks

Special concrete transport trucks (**in–transit mixers**) are made to transport and mix concrete up to the construction site. (They can be charged with dry materials and water, with the mixing occurring during transport. They can also be loaded from a "central mix" plant, with this process the material has already been mixed prior to loading.) The concrete mixing transport truck maintains the material's liquid state through agitation, or turning of the drum, until delivery. The interior of the drum on a concrete mixing truck is fitted with a spiral blade. In one rotational direction, the concrete is pushed deeper into the drum. This is the direction the drum is rotated while the concrete is being transported to the building site. This is known as "charging" the mixer. When the drum rotates in the other direction, the Archimedes' screw-type arrangement "discharges", or forces the concrete out of the drum. From there it may go onto chutes to guide the viscous concrete directly to the job site. If the truck cannot get close enough to the site to use the chutes, the concrete may be discharged into a concrete pump, connected to a flexible hose, or onto a conveyor belt which can be extended for some distance (typically ten or more metres). A pump provides the means to move the material to precise locations, multi-floor buildings, and other distance prohibitive locations. Buckets suspended from cranes are also used to place the concrete. The drum is traditionally made of steel but on some newer trucks as a weight reduction measure, fibreglass has been used.

"Rear discharge" trucks require both a driver and a "chuteman" to guide the truck and chute back and forth to place concrete in the manner suitable to the contractor. Newer "front discharge" trucks have controls inside the cab of the truck to allow the driver to move the chute in all directions.

Concrete mixers generally do not travel far from their plant, as the concrete begins to set as soon as it is in the truck. Many contractors require that the concrete be in place within 90 minutes after loading. If the truck breaks down or for some other reason the concrete hardens in the truck, workers may need to enter the barrel with jackhammers.

Workability: The workability that is required, depends primarily on how the concrete is to be placed. Concrete can be poured, pumped, and even sprayed into place, and this will affect the workability that is needed. Other factors such as the shape of the moulds, the rebar spacing, and the equipment available at the site for consolidating the fresh concrete after it is placed must also be considered. Workability is usually defined by the *slump*, which is the tendency for the fresh concrete tends to spread out under its own weight when placed onto a flat surface. Workability needs to be checked for every lot/ batch before it is poured.

9.5   Placing

Once the concrete has been adequately mixed, it must be placed into the formwork that defines its final position and shape. If the concrete is to be reinforced, the rebar must already be in place so the concrete can flow around it.

If the concrete mixing truck can be located close to (and higher than) the site, then the concrete can be poured directly into the forms. In cases where this is not possible, the concrete can be transferred in buckets by a crane or by wheelbarrow. When this is impractical due to the distance required or the size of the job, the fresh concrete can be pumped through a system of pipes or hoses to the site by special concrete pumps. Concrete that is to be pumped has more stringent requirements for workability. If the concrete is too dry, it will not pump well, while if it is too wet it will tend to segregate. Segregation can also occur if the concrete falls into the formwork too quickly, as larger aggregate particles will tend to be driven downward.

A **concrete pump** is a machine used for transferring liquid concrete by pumping. There are two types of concrete pumps.

> The first type of concrete pump is attached to a truck. It is known as a trailer-mounted boom concrete pump because it uses a remote-controlled articulating robotic arm (called a *boom*) to place concrete accurately. Boom pumps are used on most of the larger construction projects as they are capable of pumping at very high volumes and because of the labour saving nature of the placing boom. They are a revolutionary alternative to truck-mounted concrete pumps.

> The second main type of concrete pump is either mounted on a truck and known as a truck-mounted concrete pump or placed on a trailer, and it is commonly referred to as a *line pump* or trailer-mounted concrete pump. This pump requires steel or flexible concrete placing hoses/ pipes to be manually attached to the outlet of the machine. Those hoses/ pipes are linked together and lead to wherever the concrete needs to be placed.

9.6   Consolidation

Once the concrete is in place, it should be consolidated to remove large air voids developed during placement and to make sure that the concrete has flowed into all of the corners and nooks of the formwork. This process is also called compacting. Over-consolidation can lead to segregation and bleeding, but under-consolidation is more common, resulting in less-than optimal properties. The two most common methods of consolidation is vibration and roller compacting. Vibration is a mechanical process that transfers pulses of shear energy to the concrete, usually by a probe that is inserted several inches into the concrete. Each pulse of shear energy momentarily liquefies the concrete, allowing it to flow very freely. This is the standard consolidating method for general construction projects with the exception of roads. The shear energy will only travel through a limited thickness of concrete, so when a thick concrete structure is being placed the fresh concrete is poured in layers, with each layer consolidated before the next is poured over it. Vibration is a noisy and labour-intensive step, requiring expensive and specialized equipment. For this reason, there is growing use of self-consolidating concrete which flows so freely (through the use of chemical admixtures) that mechanical consolidation is not needed.

9.7   Finishing

For concrete floors and pavements, the appearance, smoothness, and durability of the surface is particularly important. Finishing refers to any final treatment of the concrete surface after it has been consolidated to achieve the desired properties. This can be as simple as pushing a wide blade over the fresh concrete surface to make it flat (screening). Floating and trowelling is a process of compacting and smoothing the surface which is performed as the concrete is starting to harden. This would be standard procedure for driveways and sidewalks. After concrete has hardened, mechanical finishing can be used to roughen the surface to make it less slippery or to polish the surface as a decorative step to bring out the beauty of a special aggregate such as marble chips. A recently developed process which is growing in popularity involves the use of concrete dyes and surface moulds to emulate the appearance of bricks, decorative pavers, or even ceramic tile. When done properly, this type of decorative concrete is almost indistinguishable from the real thing.

▪ Curing

Once concrete has been placed and consolidated, it must be allowed to cure properly to develop good final properties.

As the concrete hardens and gains strength it becomes less and less vulnerable, so the critical time period is the first hours and days after it is placed. Proper curing of concrete generally comes down to two factors, keeping it moist and keeping it supported. Hydration of cement, as the word itself implies, involves reaction with water. To cure properly, the cement paste must be fully saturated with water. If the relative humidity level inside the concrete drops to near 90% the hydration reactions will slow, and by 80% they will stop altogether. Not only will this prevent the concrete from gaining its full strength, but it will also generate internal stresses that can cause cracking. To keep fresh and young concrete moist, it can be covered with plastic or damp fabric to prevent evaporation, or sprayed periodically with water. Spraying is particularly helpful when the w/c of the concrete is low, because the original mix water is not enough for the cement to hydrate fully. The additional water will not penetrate through a thick concrete structure, but it will help create a stronger surface layer. Pools of water should not be allowed to form on the surface, however, as this will leach and degrade the concrete underneath.

When concrete is placed using formwork, there is generally a desire to remove the formwork as quickly as possible to continue the construction process. However, if this is done too soon the fresh concrete will deform under its own weight. This will lead to a loss of dimensional tolerances, cracking, or even a complete collapse. Similar problems occur if loads are applied to the surface of a floor or slab too early.

The weather plays an important role in the curing process. Hot windy weather leads to rapid evaporation and thus particular care must be taken to keep the concrete moist. Cold weather causes the concrete to harden much more slowly than hot weather. This delays the construction process, but leads to better concrete in the long run, because the hydration products develop differently at different temperatures. If fresh concrete freezes, however, it will likely be destroyed beyond repair.

- **Filling and consolidation – This activity is used for various works**

  - For creating a bund
  - For road making
  - For Storage yard making
  - For backfilling the area around foundations/ sub surface works after concreting is completed.

Compaction is the process by which the bulk density of an aggregate of matter is increased by driving out air. For any soil, for a given amount of compactive effort, the density obtained depends on the moisture content. At very high moisture contents, the maximum dry density is achieved when the soil is compacted to nearly saturation, where (almost) all the air is driven out. At low moisture contents, the soil particles interfere with each other; addition of some moisture will allow greater bulk densities, with a peak density where this effect begins to be counteracted by the saturation of the soil.

The material used in above activities plays an important role. For creating a bund, binding properties of soil are important. In road/ store yard making initial fill would be with boulders of specified size. For backfilling confined areas sand is used. Filling needs to be done in layers as specified in contract. Each layer is compacted by rollers or compactors and spraying water intermittently to achieve the specified Proctor density values. Big size boulders cannot be used for filling as cavities can exist around them. Bigger size boulders need to be broken down to acceptable size for backfilling.

- **Brick work / Masonry work –**

Brickwork is masonry produced by a bricklayer, using bricks and mortar. Typically, rows of bricks — called *courses* — are laid on top of one another to build up a structure such as a brick wall.

Constituents of brickwork are bricks, beds and perpend. The bed is the mortar upon which a brick is laid. A perpend is a vertical joint between any two bricks and is usually — but not always — filled with mortar. The dimensions of these parts are, in general, co-ordinated so that two bricks laid side by side separated only by the width of a perpend have a total width identical to the length of a single brick laid transversely on top of them.

A nearly universal rule allowing for brickwork to be stable under even modest loads is that perpends should not vertically align in any two successive courses. If this rule is observed, then the force acting on any brick is distributed across a wider area in the next successive course.

9.8   Classification of Bricks - Burnt Clay Bricks

Burnt clay bricks are commonly used bricks in the construction purpose. Burnt bricks are usually prepared using casting mould in which clay is filled through an extrusion and dried and finally baked or fired in the kiln. After they are baked in the kiln clay changes its colour or sometimes they are coloured with red paint. This is what we call burnt clay brick.

9.9   Classification of Bricks - Sand Lime Bricks

Sand lime bricks are generally produced by amalgamation of sand and hydrated lime. The mixture of sand and lime is pressed in the moulds and then given treatment in the high pressure steam autoclave.

9.10   Classification of Bricks - Concrete Bricks

This type of brick is a composition of cement and sand produced by pressing in the mould and treated by steam.

9.11   Classification of Bricks - Fly Ash Clay Bricks

This type of brick is produced by fly ash mixed with clay and finally obtained by treating in the thermal boilers.

9.12   Classification of Bricks - Fire Clay Bricks

Fire clay bricks are naturally produced bricks having ideal chemical composition and physical properties as they are obtained through mining the much of the earth's level. Fire clay bricks have more of metallic oxides than surface's bricks and are much useful in the construction.

9.13   Classification of Bricks - Fly Ash Bricks

This type of brick is produced by fly ash mixed with lime, gypsum and sand. Fly ash, lime sand and gypsum are manually fed into a pan mixer where water is added in the required proportion for intimate mixing. The proportion of the raw material is generally in the ratio 60-80% of fly ash 10-20% lime, 10% Gypsum and 10% sand, depending upon the quality of raw materials. After mixing, the mixture is shifted to the hydraulic/mechanical presses. The bricks are carried on wooden pellets to the open area where they are dried and water cured for 21 days.

Government of India has prohibited use of Brunt clay bricks in Project sites. Use of Fly Ash Bricks only is permitted for general masonry work.

9.14   Random Rubble masonry or RR Masonry

Random rubble masonry: The rubble masonry in which either undressed or hammer dressed stones are used is called random rubble masonry. Further random rubble masonry is also divided into the following:

- **Un-coursed random rubble masonry:** The random rubble masonry in which stones are laid without forming courses is known as un-coursed random rubble masonry. This is the roughest and cheapest type of masonry and is of varying appearance. The stones used in this masonry are of different sizes and shapes. Before lying, all projecting corners of stones are slightly knocked off. Vertical joints are not plumbed, joints are filled and flushed. Large stones are used at corners and at jambs to increase their strength. Once "through stone" is used for every square meter of the face area for joining faces and backing. **Suitability:** Used for construction of walls of low height in case of ordinary buildings.

- **Coursed random rubble masonry:** The random rubble masonry in which stones are laid in layers of equal height is called coursed random rubble masonry. In this masonry, the stones are laid in somewhat level courses. Headers of one coursed height are placed at certain intervals. The stones are hammer dressed. **Suitability**: Used for construction of residential buildings, go downs, boundary walls, etc.

9.15   Finish for Brickwork

The Brickwork is finished by following methods:

➢   Pointing

➢   Plastering

9.16   Pointing

The process of placing a masonry unit in a mortar bed causes the mortar to be extruded between units. This excess mortar is cleared and shaped before the mortar hardens. In some cases, the bedding mortar is meant to be seen and the mortar joint is immediately given a finish profile. In other cases, especially where the mortar is pigmented, the bedding mortar is removed from the joint before it hardens (usually a depth that averages about one inch), and then pointing mortar is applied in the void and shaped with a finish profile.

Pointing is a cheaper process for finishing brickwork. It ensures leak proof joints.

9.17   Plastering

Plastering is the finishing activity for brickwork or concrete surfaces. It is done after curing the brickwork/ concrete surface. This can be either External or Internal plastering. Cement and fine aggregate mix ratio should be 1:6 (1 Cement: 6 fine aggregate) for internal plastering of bricks and 1:4 for external plastering. Never do a plastering beyond 12 or 15mm thickness on a brick wall. Avoid plastering beyond 6 mm thickness on concrete in one go. Hacking should be done before plastering work on ceilings and columns only. Depending on the surface finish requirement, after internal plastering suitable coating is carried out.

> ■   **Roof works**

Every building needs roof before it can be occupied. The roof can be RCC slab or a slanting roof made of steel structure covered by sheets/ tiles. Before the advent of RCC, many old buildings have flat slabs with earthen floor covered with slabs supported on either wooden or steel structure.

Considering the area to be covered, many buildings prefer to have roof sheeting as a cost effective solution. For this purpose, roof trusses and purlins made of steel are used. Depending on size and load to be carried, these may be fabricated from Pipes/ Tubes or Beams, channels and angles. Instead of sheeting, in some buildings light weight pre-cast concrete slabs are also used with 40mm screed layer with tar felting on top to make it water proof.

Roof extractors are provided in all large size industrial buildings. For this, suitable size openings with matching fixing frames are provided. If roof extractors are not provided, ventilators on the side walls are provided. It is essential to ensure that there is no water ingress from these openings.

For roof sheeting, corrugated sheets are used. These can be GI sheets, coloured Steel/ Aluminium sheets or PVC/ Fibreglass sheets. For natural light to come inside building Perspex sheets are used. Sheet thickness plays an important role in durability and normally 1mm thickness is specified. For temporary structures, lesser thickness sheets are sometimes used but in case of heavy winds these sheets do not provide protection.

It is essential to ensure proper grouting/ fixing of roof structure with the basic building structure. In case of heavy winds, the roof structure can fail/ fly if not properly fixed.

Roof structure should have proper canopy and gutters to remove rain water. Roof sheets are fixed with J bolts to the structure. Proper drilling machine should be used to avoid damage to sheet and punching by nails should not be allowed. Proper size washers along with rubber seals would make the roof leak proof.

- **Cladding**

Metal cladding systems provide an efficient, attractive and reliable solution to the building envelope needs of buildings. Power plants use metal cladding for enclosing the equipment area. It is faster to erect than Brickwork. It can be only sheet or depending on requirement can be of composite section.

The roof and wall cladding, whose functions include some or all of the following:

- ➢ Separating the enclosed space from the external environment

- ➢ Transferring load to the secondary steelwork

- ➢ Restraining the secondary steelwork

- ➢ Providing thermal insulation

- ➢ Providing acoustic insulation

- ➢ Preventing fire spread

- ➢ Providing an airtight envelope

- ➢ Providing ventilation to a building (ventilated or unventilated roofs and walls).

The cladding will also normally include ancillary components such as windows, roof lights, vents and gutters.

Single skin sheeting is widely used in industrial structures where no insulation is required. The sheeting is fixed directly to the purlins or side rails as shown. The cladding is generally made from 0.7mm gauge pre-coated steel with a 32mm to 35mm trapezoidal profile depth.

Cladding sheets come in different colours thereby do not need painting.

■ **Plumbing**

Plumbing is the system of pipes, drains fittings, valves, valve assemblies, and devices installed in a building for the distribution of water for drinking, heating and washing, and the removal of waterborne wastes, and the skilled trade of working with pipes, tubing and plumbing fixtures in such systems.

Present-day water-supply systems use a network of high-pressure pumps and pipes in buildings; are now made of copper, brass, plastic (particularly cross-linked polyethylene called PEX, or other nontoxic material.) Due to its toxicity, lead is not used in modern water-supply piping. Drain and vent lines are made of plastic, steel or cast-iron.

Pipe normally has thicker walls and may be threaded or welded, while tubing is thinner-walled and requires special joining techniques such as brazing, compression fitting, crimping, or for plastics, solvent welding.

In addition to lengths of pipe or tubing, pipe fittings are used in plumbing systems, such as valves, elbows, tees, and unions. Pipe and fittings are held in place with pipe hangers and strapping.

Plumbing fixtures are exchangeable devices using water that can be connected to a building's plumbing system. They are considered to be "fixtures", in that they are semi-permanent parts of buildings, not usually owned or maintained separately. Plumbing fixtures are seen by and designed for the end-users. Some examples of fixtures include water closets (also known as toilets), urinals, bidets, showers, bathtubs, utility and kitchen sinks, drinking fountains, ice makers, humidifiers, air washers, fountains, and eye wash stations.

Threaded pipe joints are sealed with thread seal tape or pipe dope. Many plumbing fixtures are sealed to their mounting surfaces with plumber's putty.

Plumbing equipment includes devices such as water meters, pumps, expansion tanks, backflow arrestors, water filters, UV sterilization lights, water softeners, water heaters, heat exchangers, gauges, and control systems.

Specialized plumbing tools include pipe wrenches, flaring pliers, pipe vice, pipe bending machine, pipe cutter, dies and joining tools such as soldering torches and crimp tools.

- **Soak Pit**

A soak pit, also known as a soakaway or leach pit, is a covered, porous-walled chamber that allows water to slowly soak into the ground. Pre-settled effluent from a Collection and Storage/Treatment or (Semi-) Centralized Treatment technology is discharged to the underground chamber from which it infiltrates into the surrounding soil.

As wastewater (greywater or blackwater after primary treatment) percolates through the soil from the soak pit, small particles are filtered out by the soil matrix and organics are digested by microorganisms. Thus, soak pits are best suited for soil with good absorptive properties; clay, hard packed or rocky soil is not appropriate.

9.18 Design Considerations

The soak pit should be between 1.5 and 4 m deep, but as a rule of thumb, never less than 2 m above the groundwater table. It should be located at a safe distance from a drinking water source (ideally more than 30 m). The soak pit should be kept away from high-traffic areas so that the soil above and around it is not compacted. It can be left empty and lined with a porous material to provide support and prevent collapse, or left unlined and filled with coarse rocks and gravel. The rocks and gravel will prevent the walls from collapsing, but will still provide adequate space for the wastewater. In both cases, a layer of sand and fine gravel should be spread across the bottom to help disperse the flow. To allow for future access, a removable (preferably concrete) lid should be used to seal the pit until it needs to be maintained.

In a project, if the soak pit location is not properly selected, there can be conflict with local population. The soak pits overflow in monsoon when the water table has risen. In such case, arrangement for pumping out waste water and disposing it, need to be made by engaging tankers. Disposing waste water is a major problem faced at construction sites located near villages/ towns.

▪ Septic Tank

A septic tank generally consists of a concrete or plastic tank (or sometimes more than one tank) of between 4000 and 7500 litres connected to an inlet wastewater pipe at one end and a septic drain field at the other. In general, these pipe connections are made via a T pipe, which allows liquid to enter and exit without disturbing any crust on the surface. Today, the design of the tank usually incorporates two chambers (each equipped with a manhole cover), which are separated by means of a dividing wall that has openings located about midway between the floor and roof of the tank.

Wastewater enters the first chamber of the tank, allowing solids to settle and scum to float. The settled solids are anaerobically digested, reducing the volume of solids. The liquid component flows through the dividing wall into the second chamber, where further settlement takes place, with the excess liquid then draining in a relatively clear condition from the outlet into the leach field, also referred to as a drain field or seepage field, depending upon locality. A percolation test is required to establish the porosity of the local soil conditions for the drain field design.

Waste that is not decomposed by the anaerobic digestion eventually has to be removed from the septic tank, or else the septic tank fills up and wastewater containing undecomposed material discharges directly to the drainage field. Not only is this detrimental for the environment but, if the sludge overflows the septic tank into the leach field, it may clog the leach field piping or decrease the soil porosity itself, requiring expensive repairs.

When a septic tank is emptied, the accumulated faecal sludge or septage is pumped out of the tank by using a vacuum truck. How often the septic tank has to be emptied depends on the volume of the tank relative to the input of solids, the amount of indigestible solids, and the ambient temperature (because anaerobic digestion occurs more efficiently at higher temperatures), as well as usage, system characteristics and the requirements of the relevant authority. An older system with an undersized tank that is being used by a large group will require much more frequent pumping than a new system used by only a few people. Anaerobic decomposition is rapidly restarted when the tank re-fills.

9.19 Potential problems

Like any system, a septic system requires maintenance. Although septic systems generally require less maintenance than communally connected sewage systems, the maintenance of a septic system is often the responsibility of the resident or property owner. For this reason they are sometimes perceived to require higher maintenance than other systems. Neglected or abused systems can pose the following problems:

- Excessive dumping of cooking oils and grease can cause the inlet drains to block. Oils and grease are often difficult to degrade and can cause odour problems and difficulties with the periodic emptying.

- Flushing non-biodegradable items such as cigarette butts and hygiene products such as sanitary napkins, tampons, and cotton buds/swabs will rapidly fill or clog a septic tank, so as in other systems, those materials should not be disposed of in that way.

- As with all drainage systems, the use of garbage disposals for disposal of waste food can cause a rapid overload of the system with solids and early failure.

- Certain chemicals may damage the components of a septic tank, especially pesticides, herbicides, materials with high concentrations of bleach or caustic soda (lye) or any other inorganic materials such as paints or solvents.

- Certain chemicals can kill the septic bacteria needed for the system to operate. Most notably, even very small quantities of silver nitrate will kill an entire culture.

- As with communal sewage systems, roots from trees and shrubbery growing above the tank or the drain field may clog and/or rupture them.

- Playgrounds and storage buildings may cause damage to a tank and the drainage field. In addition, covering the drainage field with an impermeable surface, such as a driveway or parking area will seriously affect its efficiency and possibly damage the tank and absorption system.

- The flushing of salted water into the septic system can lead to sodium binding in the drain field. The clay and fine silt particles bind together and effectively waterproof the leach field, rendering it ineffective.

- Like communal sewage systems, excessive water entering the system will overload it and cause it to fail. Checking for plumbing leaks and practising water conservation will help optimize the system's operation.

- Very high rainfall, rapid snowmelt, and flooding from rivers or the sea can all prevent a drain field from operating, and can cause flow to back up, interfering with the normal operation of the tank. High winter water tables can also result in groundwater flowing back into the septic tank.

- Over time, bio films develop on the pipes of the drainage field, which can lead to blockage. Such a failure can be referred to as "biomat failure".

- Septic tanks by themselves are ineffective at removing nitrogen compounds that have potential to cause algal blooms in waterways into which affected water from a septic system finds its way. This can be remedied by using a nitrogen-reducing technology, or by simply ensuring that the leach field is properly sited to prevent direct entry of effluent into bodies of water.

- **Painting**

Painting is a finishing process and undertaken to protect surface from environment. Selection of paint depends on the location of project and end use of the product. Coastal areas require special paints. Exterior surfaces are given weather proof coats. Follow painting specification in the contract. At construction sites paint is mostly applied by hand brush and spray painting seldom used.

The process of painting can be divided into three steps:

a) Pre-painting work

b) Surface preparation

c) Painting

Surface preparation plays a vital role in painting. Final paint finish and durability depend on surface preparation.

> ▪ **False Ceiling**

A false ceiling is a secondary ceiling, hung below the main (structural) ceiling. It may also be referred to as a drop ceiling, T-bar ceiling, false ceiling, suspended ceiling, grid ceiling, drop in ceiling, or drop out ceiling and is a staple of modern construction and architecture in both residential and commercial applications.

Effective building design requires balancing multiple objectives: aesthetics, acoustics, environmental factors, and integration with the building's infrastructure—not to mention cost of construction as well as long-term operation costs.

A typical dropped ceiling consists of a grid-work of metal channels in the shape of an upside-down "T", suspended on wires from the overhead structure. These channels snap together in a regularly spaced pattern of cells. Each cell is then filled with lightweight ceiling tiles or "panels" which simply drop into the grid. The primary grid types are "Standard 1" (15/16 face), Slimline (9/16" grid), and concealed grid.

False ceiling while concealing electrical conduits, fire system piping and ventilation ducts, helps in reducing the heat load for air conditioning. Also helps in improving aesthetics and illumination in rooms.

> ▪ **False flooring**

False flooring is also called raised floor. A raised floor (also raised flooring, access flooring, or raised access computer floor) provides an elevated structural floor above a solid substrate (often a concrete slab) to create a hidden void for the passage of mechanical and electrical services. Raised floors are widely used in modern office buildings, and in specialized areas such as command centers, IT data centers and computer rooms, where there is a requirement to route mechanical services and cables, wiring, and electrical supply. Such flooring can be installed at varying heights from 2 inches (51 mm) to heights above 4 feet (1,200 mm) to suit services that may be accommodated beneath. Additional structural support and lighting are often provided when a floor is raised enough for a person to crawl or even walk beneath.

This type of floor consists of a gridded metal framework or substructure of adjustable-height supports (called "pedestals") that provide support for removable (lift able) floor panels, which are usually 2×2 feet or 60×60 cm in size. The height of the legs/pedestals is dictated by the volume of cables and other services provided beneath, but typically arranged for a clearance of at least six inches or 15 cm.

The panels are normally made of steel-clad particleboard or a steel panel with a cementitious internal core, although some tiles have hollow cores. Panels may be covered with a variety of flooring finishes to suit the application, such as carpet tiles, high-pressure laminates, marble, stone, and antistatic finishes for use in computer rooms and laboratories.

▪ Flooring

Flooring is the general term for a permanent covering of a floor, or for the work of installing such a floor covering. The choice of material for floor covering is affected by factors such as cost, endurance, noise insulation, comfort and cleaning effort.

Hard flooring is a family of flooring materials that includes concrete/cement, ceramic tiles, glass tiles, and natural stone products.

Ceramic tile are clay products which are formed into thin tiles and fired. Ceramic tiles are set in beds of mortar or mastic with the joints between tiles grouted. Varieties of ceramic tiles include quarry tile, porcelain, and terracotta.

Many different natural stones are cut into a variety of sizes, shapes, and thicknesses for use as flooring. Stone flooring uses a similar installation method to ceramic tile. Slate and marble are popular types of stone flooring that requires polishing and sealing. Stone aggregates, like Terrazzo, can also used instead of raw cut stone and are available as either preformed tiles or to be constructed in-place using a cement binder.

Concrete/cement finished floor is also used for its ability to be treated for different feel and its durability.

9.20   Resilient flooring

Unlike brittle tiles made of minerals, resilient flooring is made of material that has some elasticity, giving the flooring a degree of flexibility called resilience. The flooring is available in large sheets or pre-cut tiles, and either comes with pre-applied adhesive for peel-and-stick installation or requires adhesive to be troweled on to the substrate. Resilient flooring includes many different manufactured products including linoleum, sheet vinyl, vinyl composition tile (VCT), cork (sheet or tile), and rubber. Performance surfaces used for dance or athletics are usually made of wood or resilient flooring.

9.21 Seamless chemical flooring

Many different seamless flooring materials are available. These are usually latex, polyester, urethane or epoxy compounds which are applied in liquid form to provide a completely seamless floor covering. These are usually found in wet areas such as laboratories or food processing plants. These may have granular or rubberized particles added to give better traction.

▪ Facia and other Architectural works

Every Project developer needs some unique features to his buildings and Project Control room. Architects are engaged to develop designs that are approved by the Project developer. The works associated may encompass various cross discipline works by specialists.

▪ Landscaping

Landscaping is a specialised job performed to enhance the beauty of project environs. Normally this is the last activity carried out by engaging a specialist agency.

▪ Doors, shutters and Windows

Doors of various types are used in construction. The type is selected based on the use of room. The door frames are made of wood, steel or aluminium. The size of door also depends on the equipment to be installed inside and aesthetic requirements. In some cases smaller doors are fixed after the equipment is installed. Door flaps are made of Glass, steel, wood, PVC or particle boards.

Rolling shutters are made to order and installed. These may be hand operated or motorised rolling shutters.

Windows of various types are used in construction. The type is selected based on the use of room. The Window frames are made of wood, steel or aluminium.

▪ Partitions

Various types of partitions are required to be installed. The height of panels depends on use. Panelling of partitions can be glass or plywood or special materials depending on use of room. Acoustic panels are used for conference halls.

■ **Roads**

Roads either temporary or permanent need to be constructed at any project site and maintained till work completion.

Temporary roads are constructed to facilitate construction. These are either for the entire project or approach road to a module from a project road. Normally these roads are finished with gravel topping and not asphalted as a general practice. Proper slope and camber need to be maintained so that no water logging takes place. Since asphalting is not done, continuous dust suppression efforts are required by way of water sprinkling.

Permanent roads are constructed after major equipment is erected to avoid any damage. In Petroleum Industry complexes permanent concrete roads are constructed before any equipment erection starts. Availability of good roads a is pre requisite for faster construction.

■ **Drains**

Drains either temporary or permanent need to be constructed at any project site and maintained till work completion. These cater to remove surface water from rains or discharge from equipments. Special purpose drains are required to convey effluents.

During construction phase, temporary drains are made to remove water accumulated due to curing activity or seepage from excavated area. Temporary drains are also constructed to remove rain water to avoid water logging and expedite construction activity. If the ground level of project site is lower than surrounding area then there is a major problem during rainy season. Outflow from project site needs to be regulated to avoid flooding in nearby villages.

Permanent drain system takes into account rainfall, the quantum of water to be handled and suitable outflow from project site. These drains can be either open or covered drains. The wall lining can be brick or RCC.

■ **Bridges and Culverts**

Bridges and culverts either temporary or permanent need to be constructed at any project site and maintained till work completion.

- **Tanks and Reservoirs**

Tanks and reservoirs either temporary or permanent need to be constructed at any project site and maintained till work completion.

- **Trenches**

A trench is a type of excavation or depression in the ground that is generally deeper than it is wide (as opposed to a wider gully or ditch), and narrow compared to its length (as opposed to a simple hole).

In the field of construction, trenches play a major role. They are used to place underground easily damaged and obstructive infrastructure or utilities (such as piping or cables). These are RCC structures with removable covers.

- **Gunite or Shotcrete**

A cement-sand mortar that is sprayed onto formwork, walls, or rock by a compressed air ejector giving a very dense strong concrete layer: used to repair reinforced concrete, to line tunnel walls or mine airways, etc.

A mixture of cement-sand and water sprayed on a surface under pneumatic pressure.

Shotcrete has emerged as the all- inclusive industry term to correctly describe "pneumatically applied concrete" - either by the wet or dry process. The term "Gunite" is a noun (product name) and should not be used as a verb (as in, to "gunite" something). As Per the American Shotcrete Association (ASA) the correct terminology is "shotcrete - wet mix" or "shotcrete - dry mix."

The dry mix method involves placing the dry ingredients into a hopper and then conveying them pneumatically through a hose to the nozzle. The nozzleman controls the addition of water at the nozzle. The water and the dry mixture is not completely mixed, but is completed as the mixture hits the receiving surface. This requires a skilled nozzleman, especially in the case of thick or heavily reinforced sections. Advantages of the dry mix process are that the water content can be adjusted instantaneously by the nozzleman, allowing more effective placement in overhead and vertical applications without using accelerators. The dry mix process is useful in repair applications when it is necessary to stop frequently, as the dry material is easily discharged from the hose.

Wet-mix shotcrete involves pumping of a previously prepared concrete, typically ready-mixed concrete, to the nozzle. Compressed air is introduced at the nozzle to impel the mixture onto the receiving surface. The wet-gun procedure generally produces less rebound, waste (when material falls to the floor), and dust compared to the dry-mix procedure. The greatest advantage of the wet-mix process is that larger volumes can be placed in less time.

In a Power plant buried steel piping is given this protective coat on the outer surface.

▪ Steel Fabrication and Erection

During civil works, lots of steel fabrication is also involved.

Fabricated gates are used for project site and internal compounds. The gates are either wicker gate or sliding type. In sliding gate, the gate is supported on rails and moves with the help of rollers. Operation can be either manual or motor operated depending on the size and weight of gate.

Steel posts are used for fencing and these require to be fabricated as per requirement. Barbed wire fencing on top of compound wall needs supporting angles.

Watch towers and water tank support structure with approach ladders need to be fabricated and erected.

Pipe/ Cable rack structure also needs to be fabricated and erected.

Drain outflow and other openings in boundary wall need to be barred for preventing unauthorized entry.

▪ Piling

Pile foundations or piling is used in places where hard strata are not available for getting required soil bearing capacity. Typically pile foundations are found in coastal area, marshy lands and deserts. Small size piles are also used where opencast foundation for the design load is not economical.

There are many reasons a design engineer would recommend a deep foundation over a shallow foundation, but some of the common reasons are very large design loads, a poor soil at shallow depth, or site constraints. There are different terms used to describe different types of deep foundations including the pile (which is analogous to a pole), the pier (which is analogous to a column), drilled shafts, and caissons. Piles are generally driven into the ground in situ; other deep foundations are typically put in place using excavation and drilling.

9.22   Drilled piles

Drilled piles are also called caissons, drilled shafts, drilled piers, Cast-in-drilled-hole piles (CIDH piles) or Cast-in-Situ piles. Rotary boring techniques are used for larger diameter piles than any other piling method and permit pile construction through particularly dense or hard strata. Construction methods depend on the geology of the site. In particular, whether boring is to be undertaken in 'dry' ground conditions or through water-logged but stable strata - i.e. 'wet boring'.

For end-bearing piles, drilling continues until the borehole has extended a sufficient depth (socketing) into a sufficiently strong layer. Depending on site geology, this can be a rock layer, or hardpan, or other dense, strong layers. Both the diameter of the pile and the depth of the pile are highly specific to the ground conditions, loading conditions, and nature of the project.

9.23   Augercast pile

An Augercast pile, often known as a continuous flight auguring (CFA) pile, is formed by drilling into the ground with a hollow stemmed continuous flight auger to the required depth or degree of resistance. No casing is required. A cement grout mix is then pumped down the stem of the auger. While the cement grout is pumped, the auger is slowly withdrawn, conveying the soil upward along the flights. A shaft of fluid cement grout is formed to ground level. Reinforcement can be installed. Recent innovations in addition to stringent quality control allow reinforcing cages to be placed up to the full length of a pile when required.

9.24   Driven foundations

Prefabricated piles are driven into the ground using a pile driver. Driven piles are either reinforced concrete, or steel. Concrete piles are available in square, octagonal, and round cross-sections (like Franki Piles). They are reinforced with rebar and are often prestressed. Steel piles are either pipe piles or some sort of beam section (like an H-pile). Splices are used to join multiple segments end-to-end when the driven depth required was too long for a single pile; today, splicing is common with steel piles, though concrete piles can be spliced with mechanical and other means. Driving piles, as opposed to drilling shafts, is advantageous because the soil displaced by driving the piles compresses the surrounding soil, causing greater friction against the sides of the piles, thus increasing their load-bearing capacity.

9.25 Pile foundation systems

Foundations relying on driven piles often have groups of piles connected by a **pile cap** (a large concrete block into which the heads of the piles are embedded) to distribute loads which are larger than one pile can bear. Pile caps and isolated piles are typically connected with grade beams to tie the foundation elements together; lighter structural elements bear on the grade beams, while heavier elements bear directly on the pile cap.

9.26 Descriptive Questions

1. Describe the various civil works activities in a project.

2. What is a soak pit? What are design considerations for sizing and locating a soak pit?

3. What is design mix? Why it is to be made for each project?

4. What are the precautions to be taken for excavation work?

5. Describe various types of paints and their application used in civil works.

9.27 Multiple Choice Questions

1. When work is done at site by deviating from drawing/ document with approval of engineering

   i.   No action needs to be taken

   ii. As Built drawing needs to be prepared and submitted

2. Concrete test cubes are made

   i.   To line pathway

   ii. To check strength of concrete as a sample

   iii. None of the above

State True or False

SL	Statement	True	False
01	Blasting is extensively used in excavation		
02	Any water can be used for preparing concrete		
03	Sump pit needs to be provided in excavated pit		
04	No Personnel Protection equipment is required in concreting		
05	Bench marks are not essential		

UNIT - 10

CONSTRUCTION – ERECTION ACTIVITIES

10.1 Taking over foundations and area

This is the first activity in erection. The foundations need to be checked with respect to Mechanical/ Electrical drawings. Foundation bolts should be straight and concrete levels should be as per requirement. Embedments should be clear of any excess civil material and soundly embedded. Pockets in foundation should be as per drawing and clear of all debris. Orientation and size of openings should be as per drawing. Centre lines of foundations are checked and once all deviations are cleared, the foundations are taken over for further erection activity.

The following points are to be taken care during inspection.

➢ Checking of the foundations with permanent bench mark.

➢ Elevation of the foundation top surface of concrete (TOC).

➢ Centre line distance between the individual foundations.

➢ Foundation bolt diagonal distances.

➢ Diagonal distances of the individual foundations.

➢ Straightness/Verticality of the foundation bolts.

The area around foundations needs to be level, compacted and free of debris. This is required for safe movement of men, machines and material. Adequate area for unloading material and preparatory work is required near the foundations.

10.2   Foundation preparation

The tolerances in Civil construction are liberal compared to Mechanical/ Electrical erection. This necessitates additional preparation work on foundations by Mechanical/ Electrical erection agency. The activities are as listed below

➢ The foundation top surface chipping has to be carried out as required.

➢ The foundation pockets are to be cleaned thoroughly.

➢ The foundation bolts are to be cleaned & apply oil or grease over threaded portion.

➢ Ensure the availability of the packer plates/shims to the specified grout thickness.

➢ The packer plates are to be placed; grouted and top elevation has to verified and recorded.

➢ The foundation bolts are to be preserved till the erection of the main columns/ equipment.

Typical tolerances are

Deviation from major dimensions of project drawing	1mm/meter, max +15mm
Gap between base plate bottom to pedestal top (Grouting space)	+ 30mm min, 70mm max
Foundation bolt top level elevation	+10mm, -00mm

Top level of foundation	+10mm, -15mm
Centre to Centre of column pedestals	1mm/metre, max +/- 8mm
First and last row pedestals dimension deviation	1mm/metre, +/- 15mm
Centre to Centre of bolt holes in foundation	+/- 3mm
Diagonal Difference	+/- 15mm

Rotating machinery needs good contact with foundations. For this, blue matching is carried out by scraping the mating surfaces and achieving the desired contact level which can be 90% contact.

10.3   Preliminary works

These include fabrication and attachment of temporary structures like monkey ladders, working platforms and stools to permanent members for ease of erection. Care needs to be taken in attaching these temporary structures, avoiding any damage to the parent material of permanent members. At detail engineering stage, a dialogue with construction can help in providing necessary drilled holes in the basic design of the members to fix temporary structures.

10.4   Preassembly

Equipment and components are dispatched from manufacturing units in dimensions that can be easily and economically transported, safely to the project site. To save time in lifting and also avoid joints at elevations where proper working platforms cannot be provided, preassembly of components is carried out at site. Preassembly checks are carried out for critical components where deviations in tolerance are difficult to be corrected after placement.

For this purpose, preassembly beds are made of temporary steel structure and concrete blocks. The height of the bed is such that workers can work in standing position without straining. For pipes, stands with roller are used so that pipes can be rotated easily.

Typical examples of preassembly are stairs with handrails and intermediate platforms, Downcomer pipes, pressure part coils and various ducts.

10.5   Pre erection activities

This involves marking centre lines, chipping the foundations, cleaning foundation pockets, cleaning the inserts and blue matching inserts to get uniform level surface, bending projecting rebars to make space for inserting base plate of column. It also involves making barricades, working platforms and approach ladders for safe working. Clearing the area and making crane approach is part of this activity. Preassembly bed is prepared for checking the components before erection as well as preassembly.

Before erection, the parts need to be checked and prepared for lifting. This involves removing temporary transport attachments painted in yellow. Mating surfaces need to be cleaned, deburred and blue matched as required.

For column erection, temporary ladders need to be attached to the column for reaching the column top after erection.

10.6   Gas Cutting and welding

These are processes that use fuel gases and oxygen to weld and cut metals, respectively. French engineers

Edmond Fouché and Charles Picard became the first to develop oxygen-acetylene welding in 1903. Pure oxygen, instead of air, is used to increase the flame temperature to allow localized melting of the workpiece material (e.g. steel) in a room environment. A common propane/air flame burns at about 2,000°C, a propane/oxygen flame burns at about 2,500°C, and an acetylene/oxygen flame burns at about 3,500°C.

In **Gas welding**, a welding torch is used to weld metals. When two pieces of metal are heated to a temperature that produces a shared pool of molten metal, welding of metal takes place. The molten pool is generally supplied with additional metal called filler. Filler material depends upon the metals to be welded.

In **Gas cutting**, a torch is used to heat metal to its kindling temperature. A stream of oxygen is then trained on the metal, burning it into a metal oxide that flows out of the kerf as slag. While Gas welding is a manual operation, Gas cutting can be manual or semi-automatic operation.

The apparatus used in gas welding/ cutting consists basically of an oxygen source and a fuel gas source (usually cylinders), two pressure regulators and two flexible hoses (one of each for each cylinder), and a torch. This sort of torch can also be used for soldering and brazing. The cylinders are often carried in a special wheeled trolley.

The regulator is used to control pressure from the tanks to the required pressure in the hose. The flow rate is then adjusted by the operator using needle valves on the torch. Accurate flow control with a needle valve relies on a constant inlet pressure to it.

Most regulators have two stages: the first stage of the regulator is a fixed-pressure regulator whose function is to release the gas from the cylinder at a constant intermediate pressure, despite the pressure in the cylinder falling as the gas in the cylinder is used. The adjustable second stage of the regulator controls the pressure reduction from the intermediate pressure to the low outlet pressure. The regulator has two pressure gauges, one indicating cylinder pressure, and the other indicating hose pressure. The adjustment knob of the regulator is sometimes roughly calibrated for pressure, but an accurate setting requires observation of the gauge.

The hoses are specifically designed for welding and cutting metal. The hose is usually a double-hose design, meaning that there are two hoses joined together. These hoses are colour-coded for visual identification and their threaded connectors are handed to avoid accidental wrong connection: oxygen is right-handed as normal, fuel gases use a left-handed thread. These left-handed threads also have an identifying groove cut into their nuts. Colour-coding of hoses varies between countries. In India Acetylene pipeline/hose is canary yellow with dark violet band. Oxygen pipeline/hose is canary yellow with white bands.

Connections between flexible hoses and rigid fittings are made by a crimped hose clip over a barbed spigot, often referred to as 'O' clips. The use of worm-drive or Jubilee clips is specifically forbidden. The hoses should also be clipped together at intervals approximately 1 metre apart. (Not recommended for cutting applications. Because beads of molten metal given off by the process can become lodged between the hoses where they are held together, and burn through releasing the pressurised gas inside, which in the case of fuel gas usually ignites).

Acetylene is not just flammable, in certain conditions it is also an explosive. If a detonation wave enters the acetylene tank, the tank will be blown apart by the decomposition. Ordinary check valves that normally prevent back flow cannot stop a detonation wave as they are not capable of closing before the wave passes around the gate, and for that reason a flashback arrestor is needed. It is designed to operate before the detonation wave makes it from the hose side to the supply side.

Between the regulator and hose, and ideally between hose and torch on both oxygen and fuel lines, a flashback arrestor and/or non-return valve (check valve) should be installed to prevent flame or oxygen-fuel mixture being pushed back into either cylinder and damaging the equipment or making a cylinder explode. The flashback arrestor prevents the shock waves from downstream coming back up the hoses and entering the cylinder (possibly rupturing it), as there are quantities of fuel/oxygen mixtures inside parts of the equipment (specifically within the mixer and blowpipe/nozzle) that may explode if the equipment is incorrectly shut down; and acetylene decomposes at excessive pressures or temperatures. The flashback arrestor will remain switched off until someone resets it, in case the pressure wave created a leak downstream of the arrestor.

A check valve lets gas flow in one direction only. A check valve is usually a chamber containing a ball that is pressed against one end by a spring: gas flow one way pushes the ball out of the way, and no flow or flow the other way lets the spring push the ball into the inlet, blocking it. Not to be confused with a flashback arrestor, a check valve is not designed to block a shock wave. The pressure wave could occur while the ball is so far from the inlet that the pressure wave gets past before the ball reaches its off position.

10.6.1 Torch

The torch is the part that the welder holds and manipulates to make the weld. It has a connection and valve for the fuel gas and a connection and valve for the oxygen, a handle for the welder to grasp, a mixing chamber (set at an angle) where the fuel gas and oxygen mix, with a tip where the flame forms. The torch nozzle and parameters of gas should be selected based on the metal, thickness of workpiece and torch manufacturer's recommendations. Torches are also used for locally heating the metal surfaces for welding by electrical arc welding and also cleaning the surface for further work.

■ **Welding torch**

A welding torch head is used to weld metals. It can be identified by having only one or two pipes running to the nozzle and no oxygen-blast trigger and two valves knob at the bottom of the handle, letting the operator adjust the oxygen flow and fuel flow.

■ **Cutting torch**

A cutting torch head is used to cut materials. It is similar to a welding torch, but can be identified by the oxygen blow out trigger or lever.

The metal is first heated by the flame until it is cherry red. Once this temperature is attained, oxygen is supplied to the heated parts by pressing the "oxygen-blast trigger". This oxygen reacts with the metal, forming iron oxide and producing heat. It is this heat that continues the cutting process. The cutting torch only heats the metal to start the process; further heat is provided by the burning metal.

The melting point of the iron oxide is around half that of the metal; as the metal burns, it immediately turns to liquid iron oxide and flows away from the cutting zone. However, some of the iron oxide remains on the workpiece, forming a hard "slag" which can be removed by gentle tapping and/or grinding.

10.6.2   Safety Precautions in Gas welding/ cutting

Oxyacetylene welding/cutting is not difficult, but there are a good number of subtle safety points that should be learned such as:

- More than 1/7 the capacity of the cylinder should not be used per hour. This causes the acetone inside the acetylene cylinder to come out of the cylinder and contaminate the hose and possibly the torch.

- Acetylene is dangerous above 1 atm (15 psi) pressure. It is unstable and explosively decomposes.

- Proper ventilation when welding, will help to avoid large chemical exposure.

- Proper protection such as welding goggles should be worn at all times, including to protect the eyes against glare and flying sparks. Special safety eyewear must be used—both to protect the welder and to provide a clear view through the yellow-orange flare given off by the incandescing flux.

- Fuel gases that are denser than air (Propane, Propylene, MAPP, Butane, etc...), may collect in low areas if allowed to escape. To avoid an ignition hazard, special care should be taken when using these gases over areas such as basements, sinks, storm drains, etc. In addition, leaking fittings may catch fire during use and pose a risk to personnel as well as property.

- When using fuel and oxygen tanks they should be fastened securely, upright to a wall or a post or a portable cart. An oxygen tank is especially dangerous for the reason that the oxygen is at a pressure of 21 MPa when full, and if the tank falls over and its valve strikes something and is knocked off, the tank will effectively become an extremely deadly flying missile propelled by the compressed oxygen, capable of even breaking through a brick wall. For this reason, never move an oxygen tank around without its valve cap screwed in place.

- On an oxyacetylene torch system there will be three types of valves, the tank valve, the regulator valve, and the torch valve. There will be a set of these three valves for each gas. The gas in the tanks or cylinders is at high pressure.

- A less obvious hazard of welding is exposure to harmful chemicals. Exposure to certain metals, metal oxides, or carbon monoxide can often lead to severe medical conditions. Damaging chemicals can be produced from the fuel, from the work-piece, or from a protective coating on the work-piece. By increasing ventilation around the welding environment, the welders will have much less exposure to harmful chemicals from any source.

10.7 Placement

Placement involves lifting of components and placing them in desired location. For this, Lifting equipment, slings and various types of ropes are used.

- ■ **Hitches**

How wire rope slings are configured to lift a load is called a hitch. Most lifts use one of three basic hitches.

- ■ **Vertical Eye and Eye Hitch**

If one eye of the sling is attached to the lifting hook and the other eye is attached to the load, this is called a vertical eye and eye, or straight, hitch. A tagline should be used to prevent load rotation that may damage the sling. When two or more slings are attached to the same lifting hook, the total hitch becomes, in effect, a lifting bridle and the load is distributed equally among the individual slings. Slings used at an angle have a lower rated capacity than one used vertically.

- ■ **Choker Hitch**

In the choker hitch, one eye of the sling is attached to the lifting hook, while the sling itself is drawn through the other eye. The load is placed inside the "choke" that is created while the sling is drawn tight over the load through the eye.

Choker hitches reduce the lifting capability of a sling since the wire rope component's ability to adjust during the lift is affected. You should only use a choker hitch when the load will not be seriously damaged by the sling body, or the sling damaged by the load, and when the lift requires the sling to hug the load. Never choke a load so that any part of one eye or splice is in the part of the sling that passes through the other eye to form the choke.

Two notes of caution: Always pull a choker hitch tight before the lift is made. It should never be pulled down during the lift. Also, never use only one choker hitch to lift a load that could shift or slide out of the choke.

- ■ **Basket Hitch**

A basket hitch is formed when both eyes of the sling are placed on the lifting hook, thereby forming a circular basket of the sling. This type of hitch distributes the load equally between the two legs of the sling, within limitations.

▪ Lifting Bridles

When you attach two or more slings to the same lifting hook, or are connected to a link rigged onto the hook, the total hitch becomes a lifting bridle, distributing the load among the individual slings. When using two or more slings as a lifting bridle, remember that the sling angle affects the slings' rated capacities. Also, the location of the lift's centre of gravity will affect the load on each sling leg.

There are four primary factors to take into consideration when lifting a load. They are:

(1) The physical parameters of the load;

(2) The number of legs and the angle they make with the horizontal;

(3) The rated capacity of the sling; and

(4) The condition of the sling.

▪ Physical parameters of the load

The size of the object to be lifted, and particularly the location of lifting points, will affect sling selection.

The weight of the lift, while a critical component, is only a part of the information. The location of the centre of gravity is also necessary to determine sling loadings. If the load has small diameter corners, protective blocking or "softeners" must be used so that sling capacity isn't reduced. Also, if lifting a painted object or an object with a finished surface, padding or softeners may be needed between the sling and the load to protect the load.

▪ Number of legs and angle with the Horizontal

As the angle formed by the sling leg and the horizontal decreases, the rated capacity of the sling also decreases. In other words, the smaller the angle between the sling leg and the horizontal, the greater the load on the sling leg. The minimum angle allowed is 30 degrees.

▪ Rated capacity

The rated capacity of a sling must never be exceeded. The rated capacity is based both on sling fabrication components (minimum breaking force of rope used, splicing efficiency, number of parts of rope in sling and number of sling legs) and sling application components (angle of legs, type of hitch, D/d ratios, etc.)

If you are using one wire rope sling in a vertical hitch, you can utilize the full rated lifting capacity of the sling, but you must not exceed that lifting capacity.

If you are using two wire rope slings in a vertical hitch (called a 2-legged bridle hitch) in a straight lift, the load on each leg increases as the angle between the leg and the horizontal plane decreases.

Whenever you lift a load with the legs of a sling at an angle, you can calculate the actual load per leg by using the following three-step formula.

▪ Three-step formula for calculating load per sling leg

These calculations assume that the centre of gravity is equal distance from all of the lifting points, and the sling angles are the same. If not, more complicated engineering calculations are needed.

1. Divide the weight of your total load by the number of legs you are using. This gives you the load per leg if the lift were being made with all legs lifting vertically.

2. Measure the angle between the legs of the sling and the horizontal plane.

3. Multiply the load per leg that you calculated in step 1 by the load factor for the leg angle you are using. Use the Load factor guidelines table to determine the load factor. The result is the actual load on each leg of the sling for this lift and angle. The actual load must never exceed the sling's vertical rated capacity.

Warning: Slings shall not be used with horizontal angles less than 30°.

▪ Condition of sling

Each sling must be inspected daily. If the sling does not pass inspection (See Criteria for discarding slings), do not use.

10.8   Rigger's 10-step checklist

1. Weigh and measure

    Before you lift, be sure you know exactly how much weight you're moving, how far you have to move it and how high you must lift it. Make sure the load's weight is within the rated capacity of the sling, including consideration of sling leg angles and load's physical parameters.

2. Use the right hitch

    Decide how to connect your load to the lifting hook and how to attach the sling to the load.

3. Choose the right sling

    Each load is different. Be sure to calculate the proper rated capacity for the angles and hitch involved as well as the right type and style for the job. If D/d ratios are smaller than those indicated, the sling's rated capacity must be reduced. Choose a sling with the proper end attachments or eye protection as well as attaching hardware. Pad all corners in contact with the sling to minimize damage to the sling.

4. Inspect the sling

    Check the sling closely to be sure it is in good condition and able to make the lift. Follow all the appropriate Safety guidelines and regulations. You cannot change the length of a sling. If a different length is needed, get a sling of the required length.

5. Rig up, not down

    Always attach the sling to the load first, and then attach it to the hook.

6. Balance the load

    Always place the eye or link in the base (bowl) of the hook to prevent point loading on the hook. In a basket hitch, always balance the load to prevent slippage. The sling's legs should contain or support the load from the sides above the centre of gravity when using a basket hitch. Be certain that the slings are long enough so that the rated capacity is adequate when you consider the angle of the legs.

7.   Test the rigging

Before you make the lift, tug lightly on the rigging to be certain that blocking, sling and load protection are in place, then lift slightly off the ground and re-check the lift.

8.   Stand clear and lift

To prevent injury, move away from the areas between the sling and load and between the sling and the crane hook or hoist hook. Let the lifting device and rigging work for you. Avoid the temptation to use your muscles to prevent swinging or movement. Use a tagline or tether. Be sure to keep clear of the suspended load.

9.   Avoid shock loading

Lift slowly with a steady application or power. Don't make sudden starts or stops, either in lifting or swinging the load.

10.   Return to storage

After you're done with your lift, inspect the sling for possible damage. If damaged and not usable, destroy the sling immediately. Otherwise, return it to your sling storage rack until your next lift.

➤   Lifting Equipments

Cranes of various types, winches, chain pulley blocks, pullers are used for lifting components.

10.9   Cranes

A **crane** is a type of machine, generally equipped with a hoist, wire ropes or chains, and sheaves, that can be used both to lift and lower materials and to move them horizontally. It is mainly used for lifting heavy things and transporting them to other places. It uses one or more simple machines to create mechanical advantage and thus move loads beyond the normal capability of a human.

■ **Types of Cranes**	

➢ **Overhead crane**

An **overhead crane**, also known as a bridge crane, is a type of crane where the hook-and-line mechanism runs along a horizontal beam that runs along two widely separated rails. Often it is in a long factory building and runs along rails along the building's two long walls. It is similar to a gantry crane. Overhead cranes typically consist of either a single beam or a double beam construction. These can be built using typical steel beams or a more complex box girder type. Single bridge box girder crane are equipped with the **hoist** and system operated with a control pendant. Double Girder Bridge cranes are more typical when needing heavier capacity systems from 10 tons and above. The advantage of the box girder type configuration results in a system that has a lower deadweight yet a stronger overall system integrity. Also included would be a hoist to lift the items, the bridge, which spans the area covered by the crane, and a trolley to move along the bridge.

➢ **Mobile cranes**

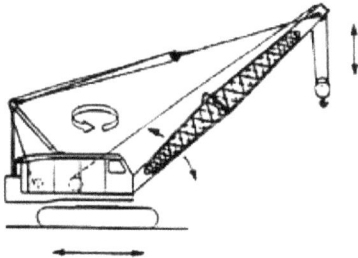

Lattice Boom Crawler Crane (LBC)   Lattice Boom Crawler Truck Crane (LBT)

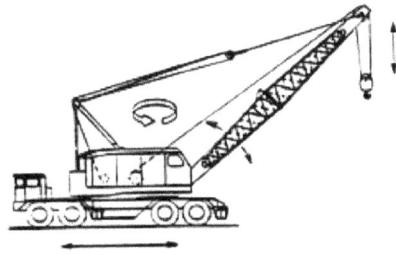

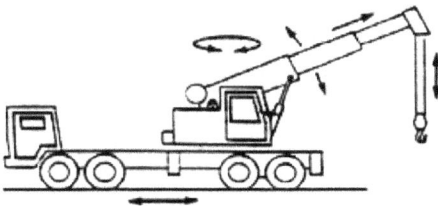

Large Telescoping Boom Crane - (Swing Cab)   Small Telescoping Boom Crane (Fixed Cab)

The most basic type of mobile crane consists of a truss or telescopic boom mounted on a mobile platform — be it on road, rail or water. Common terminology is conventional and hydraulic cranes respectively.

> ➢ **Crawler crane**

A crawler is a crane mounted on an undercarriage with a set of tracks (also called crawlers) that provide stability and mobility. Crawler cranes range in lifting capacity from about 40 MT to 3,000 MT.

Crawler cranes have both advantages and disadvantages depending on their use. Their main advantage is that they can move around on site and perform each lift with little set-up, since the crane is stable on its tracks with no outriggers. In addition, a crawler crane is capable of travelling with a load. The main disadvantage is that they are very heavy, and cannot easily be moved from one job site to another without significant expense. Typically a large crawler must be disassembled and moved by trucks, rail cars or ships to its next location.

> ➢ **Pick and carry crane**

A Pick and Carry Crane is similar to a mobile crane in that is designed to travel on public roads, however Pick and Carry cranes have no stabiliser legs or outriggers and are designed to lift the load and carry it to its destination, within a small radius, then be able to drive to the next job. The capacity range is usually ten to twenty tonnes maximum lift, although this is much less at the tip of the boom. Pick and Carry cranes have displaced the work usually completed by smaller truck cranes as the set up time is much quicker. Many steel fabrication yards also use pick and carry cranes as they can "walk" with fabricated steel sections and place these where required with relative ease.

> ➢ **Truck-mounted crane**

A crane mounted on a truck carrier provides the mobility for this type of crane. This crane has two parts: the carrier often referred to as the Lower, and the lifting component which includes the boom, referred to as the Upper. These are mated together through a turntable, allowing the upper to swing from side to side. These modern hydraulic truck cranes are usually single-engine machines, with the same engine powering the undercarriage and the crane. The upper is usually powered via hydraulics run through the turntable from the pump mounted on the lower.

Generally, these cranes are able to travel on highways, eliminating the need for special equipment to transport the crane, unless weight or other size constrictions are in place, such as local laws. If this is the case, larger cranes are equipped with either special trailers to help spread the load over more axles or are able to disassemble to meet requirements. An example is counterweights. Often a crane will be followed by another truck hauling the counterweights that are removed for travel. In addition some cranes are able to remove the entire upper. However, this is usually only an issue in a large crane and mostly done with a conventional crane. When working on the job site, outriggers are extended horizontally from the chassis then vertically to level and stabilize the crane while stationary and hoisting. Many truck cranes have slow-travelling capability (a few KM per hour) while suspending a load. Great care must be taken not to swing the load sideways from the direction of travel, as most anti-tipping stability then lies in the stiffness of the chassis suspension. Most cranes of this type also have moving counterweights for stabilization beyond that provided by the outriggers. Loads suspended directly aft are the most stable, since most of the weight of the crane acts as a counterweight. Factory-calculated charts (or electronic safeguards) are used by crane operators to determine the maximum safe loads.

Truck cranes range in lifting capacity from 13 MT to 1,100 MT.

> **Rough terrain crane**

A crane mounted on an undercarriage with four rubber tires that is designed for pick-and-carry operations and for off-road and "rough terrain" applications. Outriggers are used to level and stabilize the crane for hoisting.

These telescopic cranes are single-engine machines, with the same engine powering the undercarriage and the crane, similar to a crawler crane. In a rough terrain crane, the engine is usually mounted in the undercarriage rather than in the upper, as with crawler crane. Most have 4 wheel drive and 4 wheel steering which allows them to traverse tighter and slicker terrain than a standard truck crane with less site preparation.

> **All terrain cranes**

It is a mobile crane with the necessary equipment to travel at speed on public roads, and on rough terrain at the job site, using all-wheel and crab steering. AT's combine the roadability of Truck-mounted Cranes and the manoeuvrability of Rough Terrain Cranes.

AT's have 2-9 axles and are designed for lifting loads up to 1,200 MT

> **Gantry crane**

A gantry crane has a hoist in a fixed machinery house or on a trolley that runs horizontally along rails, usually fitted on a single beam (mono-girder) or two beams (twin-girder). The crane frame is supported on a gantry system with equalized beams and wheels that run on the gantry rail, usually perpendicular to the trolley travel direction. These cranes come in all sizes, and some can move very heavy loads. At construction sites, these cranes are used in stores, preassembly yard and fabrication yard.

> **Tower crane**

Tower cranes are used extensively in construction and other industry to hoist and move materials. There are many types of tower cranes. Some Tower cranes are stationery and fixed to the ground, other type of crane moves on rails. Although they are different in type, the main parts are the same, as follows:

- **Mast**: the main supporting tower of the crane. It is made of steel trussed sections that are connected together during installation.

- **Slewing Unit**: the slewing unit sits at the top of the mast. This is the engine that enables the crane to rotate.

- **Operating Cabin**: the operating cabin sits just above the slewing unit. It contains the operating controls.

- **Jib**: the jib, or operating arm, extends horizontally from the crane. A "luffing" jib is able to move up and down; a fixed jib has a rolling trolley that runs along the underside to move goods horizontally.

- **Hook**: the hook (or hooks) is used to connect the material to the crane. It hangs at the end of thick steel cables that run along the jib to the motor.

- **Weights**: Large concrete counterweights are mounted toward the rear of the mast, to compensate for the weight of the goods lifted.

10.10   Strand jack equipment

With development of hydraulic power packs, this new system has become popular for lifting heavy loads where cranes cannot be used or use of crane is not practical. In this system the strand is a wire rope which passes through a hydraulic jack. Depending on the load to be lifted the number of strands is decided. On the structure above the final load position, temporary structure is erected on which hydraulic power pack is placed. The jack has two anchor/ grips, one at bottom and one on the top of hydraulic cylinder. By alternately holding and releasing the grips, the load is lifted with every stroke of hydraulic cylinder. For longer jobs the number of hydraulic cylinders can be increased to limit the load on individual jack.

The system is now used for lifting boiler drums, generator stator and sometimes ceiling girders. There are contractors specialising in this work who bring their own equipment.

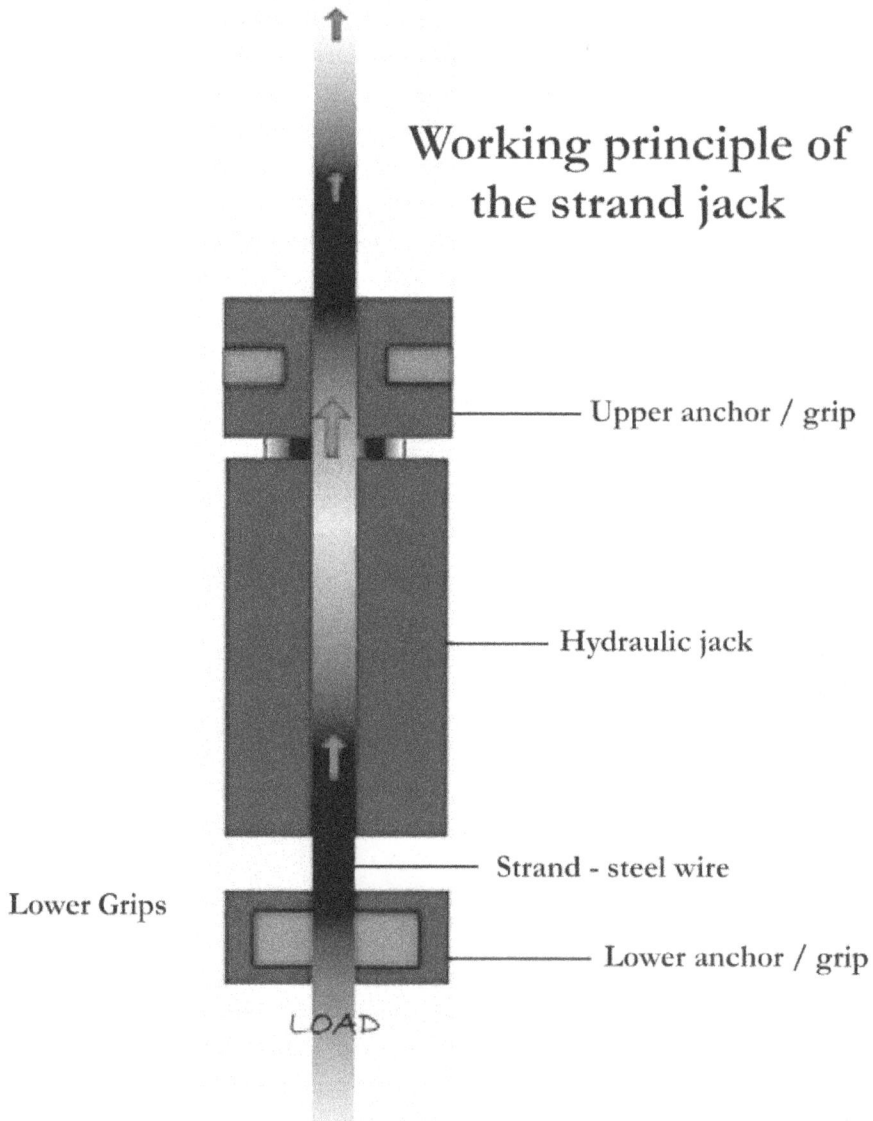

Working principle of the strand jack

Upper anchor / grip

Hydraulic jack

Strand - steel wire

Lower Grips

Lower anchor / grip

LOAD

10.11   Winch

A **winch** is a mechanical device that is used to pull in (wind up) or let out (wind out) or otherwise adjust the "tension" of a rope or wire rope (also called "cable" or "wire cable"). In its simplest form it consists of a spool and attached hand crank. In larger forms, winches stand at the heart of machines as diverse as tow trucks and elevators. The spool can also be called the winch drum. More elaborate designs have gear assemblies and can be powered by electric, hydraulic, pneumatic or internal combustion drives. Some may include a solenoid brake and/or a mechanical brake or ratchet and pawl device that prevents it from unwinding unless the pawl is retracted.

10.12   Chain Pulley Block

A chain pulley is a simple mechanical tool to make the physical act of lifting objects much simpler. Consisting of a linked chain, a metal housing and several gears, a chain pulley uses physics and geometry to increase lifting efficiency. Normally it has two sets of chains; heavier chain has a hook connected to the load while smaller chain is used to operate the mechanism.

The chain pulley uses a variety of chain sizes depending on the job it is built to perform. Thinner chains are useful in jobs that require speed and do not involve lifting massive weights. Large, thick chains, conversely, take more effort to run through the chain pulley system, but they can withstand greater loads. Much thought is put into selecting the perfect type of chain to use in different situations.

Chain pulley housing is the most visible part of the mechanism, and it varies greatly in size. Constructed of durable metals, such as steel, the housing locks the gear wheels into place. They are constructed to withstand large amounts of pressure because the point where the housing is connected to a stable position, likely a ceiling beam or crane, must endure great amounts of strain.

10.13   Slings

It is a component frequently made of wire rope or synthetic fibre used in a rigging system for lifting. The material of sling selected depends on the load to be lifted, nature of the load and chances of any damage to the load or load surface due to friction between sling and load while lifting.

Indian Standard IS 2762:2009 gives specification for wire rope slings and sling eyes.

Guidelines for use of slings is 2762: 2009

   a)   Assess load to be lifted and its position of centre of gravity.

   b)   Ensure that a proper sling is chosen for lifting the load.

c) Select the appropriate method of slinging.

d) The effective diameter of double pan should not be less than twice the diameter of rope.

e) While fitting eye termination particularly with thimble reinforcement to the lifting hook. Ensure that it is seated properly without overcrowding.

f) Slings shall not be used for bending or strapping unless they are so designed for lifting purposes.

g) Avoid use of Lang's lay rope for making sling.

h) Do not use hand spliced sling if it is likely to rotate during lifting the load.

j) Lift the load slowly avoiding jerk, shock etc.

k) Capacity of master link (ring) shall be at least equal to the capacity of the full sling. For intermediate link (ring) it shall be at least 1.4 times that of one sling leg.

i) For Two or more leg slings. Maximum permissible angle to the vertical for any sling leg shall not be more than 60°.

m) Do not use sling beyond its permissible service temperature as given below

Type of Splicing	Wire rope with	Service Temperature °C	Bearing capacity percent
Mechanically spliced with Aluminium Ferrule	Fibre core	-60 to +100	100
Steel core			
Hand spliced	Fibre core	-60 to +100	100
Hand spliced	Steel core	-60 to +250	100
+250 to +400		75	

10.13.1 Sample Capacity Chart for slings

M* = mode factor			CAPACITY					
ITEM CODE	AVG. DIA	COLOUR	S.W.L	S.W.L	S.W.L	S.W.L	S.W.L	Guaranteed Min
			Vertical lift	Basket lift	Choke r lift	45deg lift	90deg lift	Breaking load
	MM		M* = 1	M = 2	M = 08	M = 1.8	M = 1.4	
			kgs.	kgs.	kgs.	kgs.	kgs.	kgs.
XX 500	15	VOILET	500	1000	400	900	700	3500
XX 1000	18	VOILET	1000	2000	800	1800	1400	7000
XX 2000	20	GREEN	2000	4000	1600	3600	2800	14000
XX 3000	22	YELLOW	3000	6000	2400	5400	4200	21000
XX 4000	25	GREY	4000	8000	3200	7500	5600	28000
XX 5000	27	RED	5000	10000	4000	9000	7000	35000
XX 6000	32	BROWN	6000	12000	4800	10800	8400	42000
XX 8000	38	BLUE	8000	16000	6400	14400	11200	56000
XX 10000	46	ORANGE	10000	20000	8000	18000	14000	70000
XX 12000	58	ORANGE	12000	24000	9600	21600	16800	84000
XX 15000	70	ORANGE	15000	30000	12000	27000	21000	105000
XX 20000	78	ORANGE	20000	40000	16000	36000	28000	140000
XX 25000	90	ORANGE	25000	50000	20000	45000	35000	1750000
XX 30000	100	ORANGE	30000	60000	24000	54000	42000	210000
XX 36000	114	ORANGE	36000	72000	28800	64800	50400	252000
XX 40000	125	ORANGE	40000	80000	32000	72000	56000	280000
XX 50000	180	ORANGE	50000	100000	40000	90000	70000	350000
XX 75000	200	ORANGE	75000	150000	60000	1350000	105000	525000
XX 100000	250	ORANGE	100000	200000	80000	180000	140000	700000

10.13.2 Criteria for discarding slings as per IS

Presence of broken wires, excessive wear, mechanical and other damages due to heat, chemical reaction etc are the main criteria for discarding a sling during use. Appearance of any of the following kinds of damage shall be a reason to withdraw a sling from the service:

a) Broken strand.

b) Slackening under no load.

c) Crushing under no load.

d) Visible wire breakage at any point on a sling for a length of:

3 x d> four numbers

6 x d> Six numbers

30 x d> Sixteen numbers

Where d is diameter of rope.

e) Crushing at the load bearing point of the eye along with four broken wires.

f) Kink formation.

g) Sign of corrosion.

h) Damage or undue wear at the eye termination.

10.14 Accessories

This includes pulleys either single or multi sheave pulleys, D shackles, hooks, clamps etc. Selection depends on the load to be lifted and lifting equipment.

10.15 Ropes

Ropes made of different material are used in construction. These include wire ropes, manila ropes and fibre ropes to quote a few. It is to be noted that ropes should be used within their safe working load capacities only.

Wire ropes are composite construction of strands of fibre or steel or both, bunched to form a core surrounded by stranded and twisted cores. A wire rope is designated by the overall diameter, the number of strands, number of wires per strand, the cores, the type of lay and the length etc.

10.16 Surface Finish, Fits and Tolerances

In India Surface Roughness is specified as per IS 3073 (1967) and Surface texture indicated on technical drawings IS 10719 (1983). In the United States, surface finish is usually specified using the ASME Y14.36M standard. The other common standard is International Organization for Standardization (ISO) 1302.

c		d	Lay	a	Surface parameter
		‗	Parallel		D F S-L / Rz N C V
		⊥	Perpendicular		
		X	Cross-hatch	D	Tolerance direction, upper (U) or lower (L)
		M	Multi-directional	F	Filter type, for example "2RC"
		C	Circular	S	Short filter cutoff, for removing noise
b	Secondary surface parameter	R	Radial	L	Long filter cutoff, for removing waviness
c	Manufacturing method	P	Particulate	R	Profile type, primary (P), waviness (W), or roughness (R)
e	Minimum material removal			z	Parameter type, for example "a" for Ra or "3z" for R3z
				N	Assesment length; multiple of sampling length, usually 5
Material removal not allowed		Material removal required		C	Comparison rule, "max" for 100%, "16%" for 116%
				V	Specified value in micrometers

During assembly of rotating parts, Fits and Tolerances play an important role. The drawing specifies the tolerances and parts are manufactured to achieve desired fit after assembly. Based on the desired fit, necessary equipment needs to be arranged for either assembly or dismantling.

This includes oil bath with heating arrangement and bearing puller etc.

10.17 System of Fits and Tolerances

➢ **The standard reference temperature** is 20°C for industrial measurements and, consequently, for dimensions defined by the system.

➢ Due to the inevitable inaccuracy of manufacturing methods, a part cannot be made precisely to a given dimension; the difference between maximum and minimum limits of size is the **tolerance.**

➢ When two parts are to be assembled, the relation resulting from the difference between their sizes before assembly is called a **fit.**

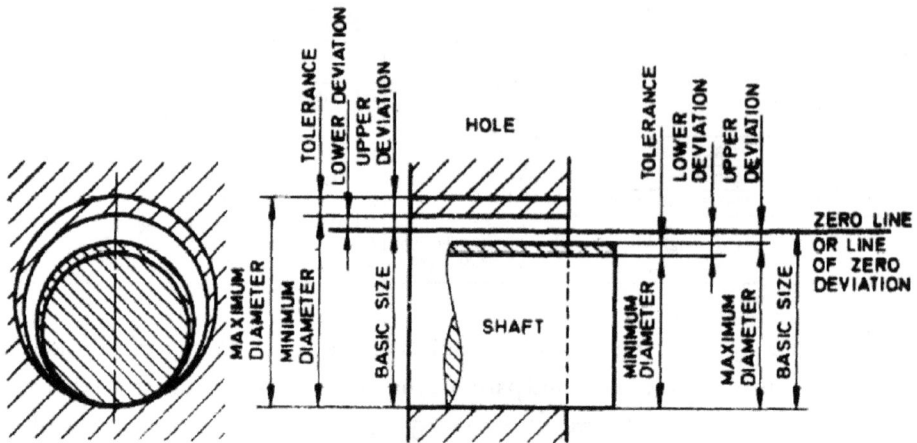

Clearance fit: In this type of fit, the largest permitted shaft diameter is less than the smallest diameter of hole so that the shaft can rotate or slide according to the purpose of the assembly.

Interference Fit: It is defined as the fit established when a negative clearance exists between the sizes of holes and the shaft. In this type of fit, the minimum permitted diameter of the shaft is larger than the maximum allowable diameter of the hole. In case of this type of fit, the members are intended to be permanently attached.

Ex: Bearing bushes, Keys & key way

Transition Fit: In this type of fit, the diameter of the largest allowable hole is greater than the smallest shaft, but the smallest hole is smaller than the largest shaft, such that a small positive or negative clearance exists between the shaft & hole.

Ex: Coupling rings, Spigot in mating holes, etc.

10.18   Alignment

After placement, the next step is alignment. Alignment can be two dimensional or three dimensional. After placement the components need to be aligned to achieve drawing dimensions and tolerances. Use of piano wires was extensive, but present day equipments have reduced this dependence.

In alignment linear dimensions and diagonal dimensions are checked for achieving specific shape. Verticality is ensured by measuring deviation from vertical in all directions. Parallelism between assemblies is also ensured by aligning them to a reference line.

Turbine shaft and casing are aligned three dimensionally to ensure concentricity and achieving linear tolerances.

During alignment, pullers may be used for bringing parts together or wedges used to match axis.

Dial gauges are used to check coupling face alignment and radial tolerances.

Laser based alignment equipments are available.

For weld joints, weld gap as per drawing is achieved and the parts are tack welded or locked in position as permitted by code. Welding is permitted only after alignment is cleared for further work.

10.19   Bolt fixing/ tightening

Bolts of various types and grades are used. Bolt material depends on working temperature and environment. Care has to be taken to identify correct grade and type of fasteners with drawing before use. The bolts are specified as per IS 1367 or ISO 898. User should get acquainted with the marking system of the standards.

During placement, dowels are used to match bolt holes and then regular bolts are fixed to avoid undue load on the bolts during erection.

For some structures, High strength friction grip (HSFG) bolts are used. Torque to be applied is specified in the drawing. The specified torque can be applied either by manual torque wrench or pneumatic torque wrench.

For heat exchangers, non elongating studs are used. The material is as per ASTM.

For Turbine Generator foundation bolts, bolt elongation is specified. To achieve this, bolt stretching device is used and then nuts tightened, to keep the bolts in tension in cold condition.

10.20   Grouting

It is a process to fill voids. In civil works it is used to fill the voids noticed and strengthen the structure. In Mechanical and Electrical works, grouting creates a bond between the foundations and the assemblies above giving them stability.

General Purpose Construction Grout are specifically designed for anchor or base plate grouting, under machinery and stanchion plates, grouting rails and bridge bearings, fixing bolts, parapet rails etc.

There are 2 types of cementitious grouts: Metallic and Non-Metallic. The metallic grouts are older technology, are prone to rusting but are good in repetitive dynamic loading.

A 3rd type of grout is the epoxy grouts which are best for the dynamic loading applications.

Surfaces for grouting must be clean and in a saturated-surface dry (SSD) condition.

Application of grout should be from one side of the area to be grouted and in one continuous flow. The use of a head box is recommended on larger applications.

It is best, when installing base plate grout, to bevel the edge of the grout @ a 45° angle from the underside of the base plate.

Always cure the exposed surfaces of the grout.

It is recommended to have a minimum annular space of 15 MM when, non-shrink cementitious grouts are used for doweling or anchoring.

After grouting, curing should be carried out in the same way as done for concrete. Further loading on structures is not permitted till the grout has achieved its strength.

10.21   Piping Erection

Piping erection is a specialised work. At project site both underground and above ground piping is carried out. It is divided into low pressure and high pressure piping.

For piping erection, General arrangement drawings at various levels, Isometric piping drawings, Hanger schedule, Valves schedule, support drawings and P&Ids are required. GA Drawings of parts where piping gets connected is also required for connecting to proper nozzles on the equipment.

Maintaining slope of piping and orientation of nozzles is very important. A wrong slope of piping can damage equipment during operation.

Depending on the pressure and medium passing through piping, piping connections are threaded, flanged, socket welded or butt welded.

Depending on the flow medium, different codes are applicable to piping fabrication and erection. Applicable code is mentioned in drawing.

Before erecting pipe pieces, these should be thoroughly cleaned internally by wire brush or mechanical means. For some piping systems, the pipes are cleaned internally by acid which is called acid pickling. Elaborate arrangements with acid tank and associated safety precautions are required to be made.

It is preferable to erect piping on permanent supports, but most of the time, supply of hangers and associated components gets delayed. In such cases piping erection is carried out on temporary supports.

Pipe supports are either fixed supports or spring supports. Spring supports also called hangers are of two types; Constant Load Hangers and Variable load hangers. These need to be locked during hydraulic test of piping to avoid damage. Steam piping spring hangers have cold setting and hot setting that differ. Initially hangers are set to cold setting value and later adjusted to hot setting when the piping is charged with steam and attains its working temperature. Some types of piping systems have bottom support springs and these also need to be set properly to avoid any displacement during operation.

Stress analysis is carried out for steam piping and supports are designed accordingly as the piping expands in all three directions. A badly designed and erected piping can damage supporting steel structure and connected equipment.

After completion and hydraulic test of piping system it is either flushed or air/ steam blown to remove all foreign material and mill scales.

10.22   Thermal Insulation

Thermal insulation is applied to equipment and piping when the differential between ambient temperature and working temperature is quite high. Thermal insulation improves the efficiency of process, reduces heat loss/ gain and also prevents accidental exposure to hot/ cold surfaces.

In power projects, air conditioning system and ducting is insulated to retain low air/ coolant temperature till the point of delivery. Boiler & Auxiliaries, Steam Turbine & Auxiliaries and steam piping is insulated to contain the heat inside the system. It is required to keep the exposed surface temperature below 65°C after insulation.

Thermal insulation is applied on surfaces after carrying out leak test or hydraulic test as the case may be.

The temperature range, within which the term "thermal insulation" will apply, is from -75°C to 815°C. All applications below -75°C are termed "cryogenic" and those above 815°C are termed "refractory"

Thermal insulation is further divided into three general application temperature ranges as follows:

A. LOW TEMPERATURE THERMAL INSULATION

  1. From 15°C through 1°C -i.e. Cold or chilled water.

  2. 0°C through -40°C - i.e. Refrigeration or glycol.

  3. -41°C through -75°C - i.e. Refrigeration or brine.

  4. -76°C through -273°C (absolute zero) - i.e. Cryogenic

B. INTERMEDIATE TEMPERATURE THERMAL INSULATION

  1. 16°C through 100°C -i.e. Hot water and steam condensate.

  2. 101°C through 315°C - i.e. Steam, high temperature hot water.

C. HIGH TEMPERATURE THERMAL INSULATION

  1. 316°C through 815°C -i.e. Turbines, breechings, stacks, exhausts, incinerators, boilers

GENERIC TYPES AND FORMS OF INSULATION

Insulation is discussed according to their generic types and forms. The type indicates composition (i.e. glass, plastic) and internal structure (i.e. cellular, fibrous). The form implies overall shape or application (i.e. board, blanket, pipe covering).

TYPES

1. Fibrous Insulation -composed of small diameter fibres which finely divide the air space. The fibres may be perpendicular or parallel to the surface being insulated, and they may or may not be bonded together. Silica, rock wool, slag wool and alumina silica fibres are used. The most widely used insulations of this type are glass fibre and mineral wool. Glass fibre and mineral wool products usually have their fibres bonded together with organic binders that supply the limited structural integrity of the products.

2. Cellular Insulation - composed of small individual cells separated from each other. The cellular material may be glass or foamed plastic such as polystyrene (closed cell), polyisocyanurate and elastomeric.

3. Granular Insulation -composed of small nodules which may contain voids or hollow spaces. It is not considered a true cellular material since gas can be transferred between the individual spaces.

This type may be produced as a loose or pourable material, or combined with a binder and fibres or undergo a chemical reaction to make a rigid insulation. Examples of these insulations are calcium silicate, expanded vermiculite, perlite, cellulose, diatomaceous earth and expanded polystyrene

FORMS

Insulations are produced in a variety of forms suitable for specific functions and applications. The combined form and type of insulation determine its proper method of installation. The forms most widely used are:

1. Rigid boards, blocks, sheets, and pre-formed shapes such as pipe insulation, curved segments, lagging etc. Cellular, granular, and fibrous insulations are produced in these forms.

2. Flexible sheets and pre-formed shapes. Cellular and fibrous insulations are produced in these forms.

3. Flexible blankets. Fibrous insulations are produced in flexible blankets.

4. Cements (insulating and finishing). Produced from fibrous and granular insulations and cement, they may be of the hydraulic setting or air drying type.

5. Foams. Poured or froth foam used to fill irregular areas and voids. Spray used for flat surfaces

MAJOR INSULATION MATERIALS

The following is a general inventory of the characteristics and properties of major insulation materials used in commercial and industrial installations. See the insulation Property Tables at the end of Section 2 for a comparative review.

CALCIUM SILICATE

Calcium silicate insulation is composed principally of hydrous calcium silicate which usually contains reinforcing fibres; it is available in moulded and rigid forms. Service temperature range covered is 35°C to 815°C. Flexural and compressive strength is good. Calcium silicate is water absorbent. However, it can be dried out without deterioration. The material is non-combustible and used primarily on hot piping and surfaces. Jacketing is field applied.

MINERAL FIBRE

a. Glass: Available as flexible blanket, rigid board, pipe covering and other pre-moulded shapes. Service temperature range is -40°C to 232°C. Fibrous glass is neutral; however, the binder may have a pH factor. The product is non-combustible and has good sound absorption qualities.

b. Rock and Slag: Rock and slag fibres are bonded together with a heat resistant binder to produce mineral fibre or wool. Upper temperature limit can reach 1035°C. The same organic binder used in the production of glass fibre products is also used in the production of most mineral fibre products. Mineral fibre products are non-combustible and have excellent fire properties.

CELLULAR GLASS

Available in board and block form capable of being fabricated into pipe covering and various shapes. Service temperature range is -273°C to 200°C and to 650°C in composite systems. Good structural strength, poor impact resistance. Material is non-combustible, non-absorptive and resistant to many chemicals.

EXPANDED SILICA OR PERLITE

Insulation material composed of natural or expanded perlite ore to form a cellular structure; material has a low shrinkage coefficient and is corrosion resistant; non-combustible, it is used in high and intermediate temperature ranges. It is available in pre-formed sections and blocks.

ELASTOMERIC FOAM

Foamed resins combined with elastomers to produce a flexible cellular material. It is available in pre-formed sections or sheets, Elastomeric insulation offer water and moisture resistance. Upper temperature limit is 105°C. Product is resilient. Fire resistance should be taken in consideration.

FOAMED PLASTIC

Insulations produced from foaming plastic resins create predominately closed cellular rigid materials. "K" values decline after initial use as the gas trapped within the cellular structure is eventually replaced by air. Check manufacturers' data. Foamed plastics are light weight with excellent cutting characteristics. The chemical content varies with each manufacturer. Available in pre-formed shapes and boards, foamed plastics are generally used in the lower intermediate and the entire low temperature ranges. Consideration should be made for fire retardancy of the material.

REFRACTORY FIBRE

Refractory Fibre insulations are mineral or ceramic fibres, including alumina and silica, bonded with extremely high temperature inorganic binders, or a mechanical interlocking of fibres eliminates the need for any binder. The material is manufactured in blanket or rigid form. Thermal shock resistance is high. Temperature limits reach 1750°C. The material is non-combustible.

INSULATING CEMENT

Insulating and finishing cements are a mixture of various insulating fibres and binders with water and cement, to form a soft plastic mass for application on irregular surfaces. Insulation values are moderate. Cements may be applied to high temperature surfaces. Finishing cements or one-coat cements are used in the lower intermediate range and as a finish to other insulation applications. Check each manufacturer's catalogue for shrinkage and adhesion properties.

PROTECTIVE COVERINGS AND FINISHES

The efficiency and service of insulation is directly dependent upon its protection from moisture entry and mechanical and chemical damage. Choices of jacketing and finish materials are based upon the mechanical, chemical, thermal and moisture conditions of the installation, as well as cost and appearance requirements. Protective coverings are divided into six functional types.

WEATHER RETARDERS

The basic function of the weather-barrier is to prevent the entry of water, ice, snow or atmospheric residue into the insulation. Sunlight and ozone can also damage certain insulations. Applications may be either jacketing of metal or plastic, or a coating of weather-barrier mastic. Jacketing must be over-lapped sufficiently to shed water. Avoid the use of plastic jacketing materials with low resistance to ultraviolet rays unless protective measures are taken.

VAPOUR RETARDERS

Vapour retarders are designed to retard (slow down) the passage of moisture vapour from one side of its surface to the other. Joints and overlaps must be sealed with a vapour tight adhesive or sealer free of pin holes or cracks. Vapour retarders take three forms:

a. Rigid jacketing -plastic fabricated jackets to the exact dimensions required and sealed vapour retarding.

b. Membrane jacketing-laminated foils, treated or coated products and plastic films which are field or factory applied to the insulation material. (Additional sealing beyond the factory seal may be necessary depending on temperature/ humidity conditions of the installation.)

c. Mastic applications -solvent types which provide a seamless coating but require time to dry.

MECHANICAL ABUSE COVERINGS

Rigid jacketing provides the strongest protection against mechanical abuse from personnel, equipment, machinery, etc. The compressive strength of the insulation material should also be considered when designing for mechanical protection.

CORROSION AND FIRE RESISTANT COVERINGS

a.   Corrosion protection -can be applied to the insulation by the use of various jacket materials. The corrosive atmosphere must be determined and a compatible material selected. Mastics may be used in atmospheres that are damaging to jacket materials'

b.   Fire resistance -can be applied to insulation systems by the use of jacketing and/or mastics. Fire resistant materials are determined by flame spread, smoke developed and combustibility. The total systems should be considered when designing for fire resistance.

APPEARANCE COVERINGS AND FINISHES

Various coatings, finishing cements, fitting covers and jackets are chosen primarily for their appearance value in exposed areas.

HYGIENIC COVERINGS

Coatings and jackets must present a smooth surface which resists fungal or bacterial growth in all areas. High temperature steam or high pressure water wash down conditions require jackets with high mechanical strength and temperature ranges.

ACCESSORIES

The term "accessories" is applied to devices or materials serving one or more of the following functions:

1.   Securement of insulation and/or jacketing

2.   Reinforcement for cement or mastic applications

3.   Stiffening around structures which may not support the weight of high density insulations

4.   Support (pipe, vessel and insulation)

5.   Sealing and caulking

6.   Water flashing

7.   Compensation for expansion/contraction of piping and vessels

Improper design or application in one or more of these accessories is a significant factor in the failure of insulation systems

Securements: As most insulations are not structural materials they must be supported, secured, fastened or bonded in place. Securements must be compatible with insulation and jacketing materials. Possible choices include:

a. Studs and pins

b. Staples, serrated fasteners, rivets and screws

c. Clips

d. Wire or straps

e. Self-adhering laps

f. Tape

g. Adhesives

h. Mastics

Ambient temperature, humidity conditions and substrate surface cleanliness affects the efficiency of tapes and adhesives and mastics on certain installations. Check the properties of temperature range and vapour permeability before choosing adhesives. And, wherever possible, use mechanical Securements.

Reinforcement for cements and mastics: Mastics and cements should be reinforced to provide mechanical strength. The following materials can be used:

a. Fibre fabrics

b. Expanded metal lath

c. Metal meshes

d. Wire netting

Compatibility of materials must be considered to prevent corrosion.

Flashing: Materials which direct the flow of liquids away from the insulation may be constructed of metal, plastic or mastic.

Stiffening: Metal lath and wire netting can be applied on high temperature surfaces before heavy density insulation is applied

Safety Precautions in Thermal Insulation works

These recommendations are applicable to all work involving fibre glass, rock wool and slag wool products.

Wear Appropriate Clothing

- Loose-fitting, long-sleeved and long-legged clothing is recommended to prevent irritation.

- A head cover is also recommended, especially when working with material overhead.

- Gloves are also recommended. Skin irritation cannot occur if there is no contact with the skin.

- Do not tape sleeves or pants at wrists or ankles.

- Remove Synthetic Vitreous Fibre (SVF) dust from the work clothes before leaving work to reduce potential for skin irritation.

Wear Appropriate Personal Protective Equipment

- To minimize upper respiratory tract irritation, measures should be taken to control the exposure. Such measures will be dictated by the work environment and may include appropriate respiratory protective equipment. See OSHA's Respiratory Protection Standard.

- When appropriate, eye protection should be worn whenever SVF Products are being handled.

- Personal protective equipment should be properly fitted and worn when required.

Removal of Fibres from the Skin and Eyes

- If fibres accumulate on the skin, do not rub or scratch. Never remove fibres from the skin by blowing with compressed air.

- If fibres are seen penetrating the skin, they may be removed by applying and then removing adhesive tape so that the fibres adhere to the tape and are pulled out of the skin.

- SVF may be deposited in the eye. If this should happen, do not rub the eyes. Flush them with water or eyewash solution (if available). Consult a physician if irritation persists.

Minimize Dust Generation

- Keep the material in its packaging as long as practical and if possible.

- Tools that generate the least amount of dust should be used. If power tools are to be used, they should be equipped with appropriate dust collection systems as necessary.

- Keep work areas clean and free of scrap SVF material.

- Do not use compressed air for clean-up unless there is no other effective method. If compressed air must be used, proper procedures and control measures must be implemented. Other workers in the immediate area must be removed or similarly protected.

- Where repair or maintenance of equipment that is either insulated with SVF or covered with settled SVF dust is necessary, clean the equipment first with HEPA vacuum equivalent (where possible) or wipe the surface clean with a wet rag to remove excess dust and loose fibres. If compressed air must be used proper procedures and control measures must be implemented. Other workers in the immediate area must be removed or similarly protected.

- Avoid unnecessary handling of scrap materials by placing them in waste disposal containers and keep equipment as close to working areas as possible to prevent the release of fibres.

Maintain Adequate Ventilation

- Unless other proper procedures and control measures have been implemented, dust collection systems should be used in manufacturing and fabrication settings where appropriate and feasible.

- Exhausted air containing SVFs should be filtered prior to recirculation into interior workspaces.

- If ventilation systems are used to capture SVFs, they should be regularly checked and maintained.

10.23   Hydraulic Testing

Systems and vessels working at pressure other than atmospheric pressure need to be tested before putting in service. It is a safety measure. Hydrostatic tests are conducted under either the industry's or the customer's specifications, or may be required by law. The system or vessel is filled with a nearly incompressible liquid- normally the working fluid.

The test pressure is as per industry practice or as defined by regulatory code. If the system or vessel is covered under a regulatory code then the hydraulic test would be witnessed by regulatory authority for giving permission to use the system or vessel.

Hydraulic test is conducted by attaching a pressurising pump to the system. Pressure gauge of adequate range and precision is fitted on the pressurising line from the pump. Before commencing filling of the system or vessel, it is important to ensure that all connections are completed, radiography plugs are installed, root valves are erected and closed; and atleast one vent valve is open to release entrapped air.

System where it is not feasible to connect a separate pressurising pump, are tested at working pressure by operating the system pumps.

For large volume systems like steam side of condenser including flash tanks, water fill test is conducted to notice any leakages.

For utility boilers, the manufacturer specifies demineralised water as fluid to be used. It is very costly and to avoid water wastage, air fill test is carried out at 7 Kg/ cm² pressure. Any deficiencies noticed are rectified and then water filled.

After attaining the test pressure, the pump is switched off and pressure drop if any, is observed for duration as specified but not less than 30 minutes. During this holding time, visual inspection is carried out to check any leakage and also any sweating on the surface. Leakages can be from weld joints or flanges, but sweating denotes problems with parent material.

Heat exchangers need to be checked on both sides after connecting with piping.

10.24   Flushing

System flushing is an important activity in construction. It ensures that the system is freed of all foreign material including grease, oil, and mill scales etc. It is carried out after hydraulic test of the system. In power plants the following is carried out

✓   Turbine oil system flushing

- ✓ Turbine control oil system flushing

- ✓ Pre boiler system alkali flushing

- ✓ Oil system flushing of Boiler fuel oil system from pump house upto Burners

- ✓ Lube oil system flushing for auxiliaries like fans, mills and Feed pumps

- ✓ Transformer oil flushing and purification – to remove moisture and other impurities from oil filled transformers.

Flushing can be done either by using pumps in the system or by using high volume flushing system skids. Normal filter elements are removed before flushing and replaced with flushing filters. Heating and cooling arrangements are made so that effective cleaning takes place.

For pre boiler system alkali flushing separate pumps need to be installed. Flushed water cannot be drained into normal drains before neutralising. Hence making neutralising arrangement is necessary before starting alkali flushing. Care needs to be taken to ensure that drained water does not go fields outside project before neutralising and sediment settlement.

After oil flushing the system needs to be drained and cleaned thoroughly before fresh oil is put in the system.

10.25   Chemical cleaning

It is a pre-commissioning activity. Acid cleaning of boiler is a process before commissioning of boiler. To avoid steam blowing, some sites HF (Hydro fluoric acid) cleaning of steam piping is carried out after erection.

10.26   Cabling

Hundreds of kilometres of cable are laid at project site. Cabling is either underground or overhead. These include

- ✓ Power Cables – both High voltage and low voltage categories.

- ✓ Instrument cables – screened, paired and prefabricated cables including computer related cables

- ✓ Special purpose heat sensing cables for use in coal conveyors

✓ Telephone cables

Cabling involves

✓ Trenches for underground cabling – Trenches are dug to a depth specified by Indian standards and cable bedding is made. After laying cable, the trenches are filled and cable markers put to indicate cable route.

✓ Fabrication and erection of cable tray supporting structure – Cable racks are fabricated and erected at site. On cable racks tray supporting structures are fixed. When cable is routed in underground trenches, tray supporting structures are fabricated and fixed on the inserts in trench wall. Cable tray supporting structures are fabricated and fixed on structural columns and beams or on inserts in walls.

✓ Cable tray erection – Cable trays are fixed on the supporting structure either horizontally or vertically. Coupler plates are used to join cable trays to each other. PVC cable trays are screwed on supporting base.

✓ Cable laying- Cable drums are supported on stands by inserting a supporting rod and cable pulled. Cables are supported on rollers to prevent damage to cables. Cable clamps are used to hold cable in position. Cable tagging is done as per standards. Before laying, cables are checked for continuity and physical damage.

✓ Cable glanding- Cable glands are used when cables enter panel or junction box. Proper drills should be used to drill holes in the plate. Gas cutting of holes is not permitted by standards. Cable armour needs to be removed for fixing gland. Cable armour should be connected to ground terminal.

✓ Cable conduiting- For illumination cabling, metallic conduit is used. In industrial building, concealed wiring is not used. These conduits are fixed on walls, ceilings, structural beams and columns.

✓ Cable entry sealing- Different types of cable entry sealings are available. This prevents entry of rats and lizards into cable space/ galleries.

✓ Termination- Depending on size and type of cables and terminal blocks, cables are terminated i.e. fixed on terminals. Ferruling is done prior to termination to identify cables. Care has to be taken while terminating, use termination detail drawing and check continuity.

✓ Cable jointing- Cables come in standard lengths in drums. It is necessary to joint cables by using jointing kits. Experienced and trained persons only should carry out cable jointing. Loop length as per standard should be kept for repairs in future.

✓ Testing- After termination, cables need to be tested to check proper connections.

10.27   Surface Preparation

For coating applications, there are two main surface contaminants that need to be removed. Mill scale is the grey flaky oxide of iron that's present on hot rolled steel. It only forms at high temperature in the hot rolling process, so is not present in cold rolled steel. For most coating processes this must be removed, otherwise the coating only adheres to the oxide, not to the steel, and adhesion is reliant on this poor bonding of steel and mill scale.

The other oxide is brown, commonly called rust, and must also be removed. The extent of rust depends on time and conditions of storage, and can represent even millimetres thick deposits. It all must come off. **Standards** ISO 8501-1 is the standard that covers blast cleaning, and it covers blast, hand flame and acid cleaning. The chart below represents the various grades.

Table of blasting qualities and their descriptions.

Standard	Method	Description of finish
Sa 1	Blast Cleaning	Poorly adhering mill scale, rust and old paint and foreign matter are removed. Well adhered contaminants remain.
Sa 2	Blast Cleaning	Most of the mill scale rust and paint etc are removed and any remaining is very well adhered.
Sa 2½	Blast Cleaning	Mill scale, rust paint and foreign matter is removed completely. Any remaining traces are visible only as

		slight stains or discoloration in the form of spots or stripes.
Sa 3	Blast Cleaning	All mill scale, rust etc is removed and the surface has a uniform white metal appearance with no shading, stripes, spots of discoloration.
St 2	Hand or Power	Poorly adhering rust, mill scale etc are removed, leaving surface contamination that

Notes

- 'Poorly adhering' is defined for mill scale as 'able to be removed by lifting with a knife blade.'

- Acid cleaning is not normally used for any other coating system than for galvanizing.

- For galvanizing, even when steel has been blast cleaned, it is always acid cleaned as well. Therefore for hot dip galvanizing, blast cleaning is rarely required, except to remove paint, severe rust, or for creating a thicker galvanized coating.

10.28 Painting

Painting is the practice of applying paint, pigment, colour or other medium to a surface (support base). Paint applications are many, but at construction sites, it is applied by brush.

The steps involved are

✓ Surface preparation – to remove all foreign material. Surface preparation is as per specifications mentioned above.

✓ Application of primer coat – Selection of primer depends on the final paint that is specified. The maximum time between final surface preparation and prime coat application inside the fabrication shop shall be 24 hours. Structural steel subjected to outdoor exposure after final surface preparation shall be prime coated within 10 hours.

✓ Final paint – This is applied in number of coats specified. The thickness of each coat is specified.

Industrial paints are pigmented liquids or powders that are used to protect and/or beautify substrates.

Industrial paints vary widely in terms of chemistry. They typically include a resin, a solvent, additives, pigments and, in some cases, a diluent. Solvent-based paints contain volatile organic compounds (VOCs) as the carrier. Water-based paints contain mostly water, but also chemicals such as glycol ethers and alcohols as the carrier. Different types of industrial paints include:

10.28.1 Types of Paints

OIL PAINT

Contains pigments usually suspended in linseed oil, a drier, and mineral spirits or other type of thinner. The linseed oil serves as the binder for the pigments, the drier controls drying time, the thinner controls the flowing qualities of the paint. As the thinner evaporates, the mixture of pigments and oil gradually dries to an elastic skin as the oil absorbs oxygen from the air or "cures". The curing action bonds a tough paint film to the applied surface. Oil paints are used inside and outside. Application of paint in sites is by either brush or spray equipment. Other methods are more suited for factory environment.

VARNISH

Varnish consists of a solution of resins in a drying oil. Varnish contains little or no pigment. It dries and hardens by evaporation of the volatile solvents, oxidation of the oil, or both. Varnish is recommended for both outdoor and indoor applications where a hard, glossy finish that is impervious to moisture is desired.

For a satin finish, the gloss varnish surface can be rubbed down with steel wool, or a satin" varnish can be used. As a floor finish, varnish provides a hard, durable film that will not greatly alter the tone of the wood.

ENAMEL

Enamel is a varnish with pigments added. Enamel has the same basic durability and toughness of a good varnish. It produces an easy-to-clean surface, and in the proper formulation, can be used for interior and exterior applications. For the highest quality interior work, an undercoat is required.

LATEX PAINT

These consist of a dispersion of fine particles of synthetic resin and pigment in water. Latex paints are quick-drying, low in odour and thinned with water. They permit the repainting and decorating of a room within a day. Because latex paints set quickly, tools, equipment and spattered areas should be cleaned promptly with warm, soapy water.

No special primer is required for interior applications except over bare metal or wood, or over highly alkaline surfaces. Spot-priming with shellac should be avoided because shiny spots will bleed through the latex film.

Exterior latex house paint can be applied directly to old painted surfaces. On new wood, it should be applied over a primer. For other surfaces, follow specific label directions.

WATER-REDUCIBLE PAINTS

This term has come into wider use in the paint business within the past few years. These products are also called "water-base" or "water-borne" paints. They include the well -known latex products, as well as products based on new synthetic polymers. While both the groups employ water as the reducing agent, the chemistry of each is different.

For example, most latex coatings dry by solvent evaporation or coalescence.

These new synthetic polymeric paints dry by a combination of solvent evaporation and chemical cross- linking. Chemical cross-linking frequently requires the blending of two materials (these products are called "two-component" coatings) and a "digestion" time before the coating can be applied.

The blending of specific materials results in chemical cross-linking and outstanding performance features, such as mar resistance, scratch resistance, washability and stain resistance.

ALKYDS

Alkyd finishes are produced in four sheens: flat, semi-gloss, low-lustre and high-gloss. Flat finishes have a velvety texture and are used to produce a rich, softly reflective surface. Alkyd flats can often be applied to painted walls and ceilings, metal, fully cured plaster, wallboard and woodwork without a priming. When required, the primer should be of a similar material. For high alkaline surfaces, an alkali-resistant primer should be used. Semi-gloss or low-lustre types add just enough sheen to woodwork and trim for contrast with flat-finished wall surfaces. Each offers great resistance to wear and washing. Low-lustre enamels are preferred in such areas as kitchens, bathrooms, nurseries and schoolrooms. Alkyd high-gloss enamels are often used for even greater serviceability and washability.

EPOXY

It is a two-part formulation which is thoroughly mixed just before use. Epoxy finishes are extremely hard and durable and excellent for demanding applications. They can be used for protecting materials such as steel, aluminium and fibre glass. The paint film dries to a brilliant gloss. The tile-like finish is smooth, easy to clean and lasts for years under the most severe conditions.

POLYESTER-EPOXY

In this, two-component materials are usually mixed prior to application. Polyester-epoxy combines the physical toughness, adhesion and chemical resistance of an epoxy with the colour retention and permanent clarity of polyester. The film is stain resistant and moisture resistant. Polyester-epoxy is available in gloss and semi-gloss sheens, and can be applied to any firm interior surface. Pot life is a full working day.

ACRYLIC-EPOXY

Chemically, acrylic-epoxy coatings provide the resistance to staining, yellowing and scuffing of acrylic resins, combined with the toughness, acid and alkali resistance of epoxies. Their performance characteristics are almost equal to those of polyester-epoxy solvent based products and their stain-resistance is superior. Acrylic-epoxy coatings are available in gloss and semi-gloss finishes -in both clear and pigmented formulations. Colorant can be added to the pigmented products to achieve hundreds of colours. Though priced higher than conventional enamels, acrylic-epoxy coatings offer superior washability, non-yellowing characteristics, and generally 3-5 times longer life, which makes them an outstanding value for interior walls continuously subjected to hard-use conditions.

POLYAMIDE-EPOXY

Tough, two-component finish with outstanding hardness, abrasion resistance, alkali and acid resistance, and adhesion when dry. Excellent as a concrete floor finish where heavy traffic wears through an alkyd finish in a short time. For exterior applications polyamide-epoxy will chalk and lose gloss on prolonged exposure; however, film integrity is not lost.

URETHANE-MODIFIED ALKYDS

One-component finishing material for outstanding abrasion resistance on wood floors, furniture, panelling, cabinets, etc. Good resistance to normal household materials such as alcohol, water, grease, etc. It may yellow to some degree with age.

ACRYLIC-URETHANE COATINGS

It is recommended for areas that demand superior chemical and stain resistance, plus colour and gloss retention. They are suitable for both interior and exterior application on properly primed steel, aluminium and masonry which are subjected to high acids and alkalinity. These products are designed to be used in commercial and industrial applications but not in homes. Acrylic-urethane coatings have high performance properties including excellent resistance to salt, steam, grease, oils, many coolants, solvents and general maintenance type machinery fluids. They also have excellent film properties and resistance to scratching, marring and chipping. The tile-like gloss and semi-gloss finishes provide superior corrosion and abrasion resistance, while maintaining excellent gloss and colour retention on exterior exposures for long periods of time. The colour and gloss retention, and chemical resistance of acrylic-urethane coatings will exceed those of conventional high performance coatings. They also dry to the touch faster than any other heavy duty topcoat in the trade sales line.

ALUMINUM PAINT

It is an all-purpose aluminium paint formulated with varnish as the vehicle for aluminium flake pigment. As the paint dries, the aluminium flakes float to the surface, providing a reflective coating. Highly resistant to weathering, it is also suitable for interior use on wood, metal or masonry. When formulated with an asphalt base, aluminium paint offers maximum adhesion and water resistance at low cost when applied to asphalt composition.

SHELLAC

It is a long-standing favourite for finishing wood floors, trim and furniture. Shellac is thinned with alcohol and should be applied in dry, warm air to avoid clouding. It dries dust-free in 15-20 minutes. Shellac can be used as a pre-staining wash coat to obtain an even stain tone on porous or soft wood such as pine. It can also be used to change the tone of an already shellacked surface by tinting it with alcohol-soluble aniline dye. Instead of re-staining, pigmented shellac, also called shellac enamel, is often used as a sealant over stained finishes for a uniform, freshly painted surface.

10.28.2 Specifications

Specifications for industrial paints include:

- **Colour - Colours** vary widely and include black, blue, brown, gray, green, orange, purple, red, white, and yellow. Primers and finishes such as flat, satin, semi-gloss, and high-gloss are available.

- **Coverage** - Coverage is the substrate area that industrial paints can cover at a specific thickness.

- **VOC content** - VOC content is limited by government regulation and represents the amount of evaporation of carbon compounds under test conditions.

- **Pot life** - Pot life is the amount of time between the mixing stage and the gel stage in which industrial paints remain usable in a pot at 25° C.

- **Viscosity** - Viscosity measures an industrial paint's resistance to flow.

- **Specific gravity** - Specific gravity is the ratio of the density of the coating to the density of the water at a specified temperature.

Features

Industrial paints are available with a variety of special features including:

- Anti-static or conductive

- Heat resistant

- Fire retardant

- Rust preventive

- Waterproof

- Water repellent

- Interior or exterior

- Suitable for touch-up coatings

- Able to withstand high temperatures

- Elastomeric - suitable for substrates that are stretched to twice their original size and then returned to their original dimensions.

- Electro-resistive - include both conductive and nonconductive materials.

Safety Precautions in Painting and Paint Handling

Be aware of these potential hazards:

- Rashes, swelling from short term skin contact

- Eye irritation, sore throat, cough, fatigue, nausea, dizziness from short term inhalation

- Liver, kidney, lung, digestive system, central nervous system damage from long term or massive exposure

- Fire -avoid using paint in an unventilated area; never expose it to an ignition source such as a spark, lit cigarette, or static electricity

- Explosion, especially if a closed container is exposed to high heat

- Reactivity from mixing with or exposure to other substances, including water

Most paint is not an especially high risk substance, but many paints contain ingredients that can cause health and safety problems. Workers must know the hazards and the basic protective measures that can make painting safe.

Paint Storage

The most important component of a storage area for paint products is a cabinet designed specifically for storing flammables. Flammables must be stored in a properly labelled flammables cabinet that has appropriate signage. In addition, follow these guidelines for your storage area:

- Set it up in an easily accessible location that is cool, dry, and well ventilated.

- Install a class-B fire extinguisher and, if metallic powders are present, a class-D fire extinguisher.

- Stock the area with appropriate clean-up equipment

Spray Booths

Spray booths require special precautions:

- Spray booths have built in ventilation systems that provide fresh air in the booth while venting out hazardous substances

- Always make sure the ventilation system is working properly before painting

- Never use anything that could spark or flame

- Space heaters, hot surfaces, portable lamps or trash that could catch fire should be kept out

- Don't keep more paint than you need for the job in the booth

- Remove debris from the booth immediately and dispose of it properly

- Have fire extinguishers and/or sprinklers nearby

- Respirators are required when working in a spray booth

Safety Precautions

Each product has specific safety precautions listed on the label. The following are some basic safety steps to keep in mind when using any paint.

- Always read and follow all the instructions and safety precautions on the label - DO NOT assume you already know how to use the product

> ➢ The label will tell you what actions to take to reduce hazards and the first aid measures to use if there is a problem

- There must be plenty of fresh air where you paint- open all doors and windows to the outside (not to hallways)

  > ➢ Place a box fan in the window, blowing out to ensure air movement

  > ➢ Continue to provide fresh air after painting - ventilation should be continued for two or three days

- Follow paint can directions for the safe cleaning of brushes and other equipment

  > ➢ Latex paint usually cleans up with soap and water

  > ➢ Oil based paint require specific products as listed on the label.

  > ➢ NEVER use gasoline to clean paint brushes -gasoline is extremely flammable.

- Buy only what you need, and store or throw away the unused amount.

- If you have leftover paint, close the container tightly

- Follow directions on the paint can on how to dispose of the product

- Flammable paint must be stored in a Flammable Liquids Storage Cabinet

- Don't mix paints with other substances without approval

- Keep paints away from ignition sources and never smoke in areas where paint is used or stored

- Know where the MSDS book is kept and how to read an MSDS. Check labels of all chemicals and MSDS's for ingredients, hazard, protective procedures and PPE.

Personal Protective Equipment

- Clothing that fully covers the skin

- Gloves that resist specific paint ingredients

- Eye/face protection if recommended

- Safety glasses, goggles, hoods or face shields

- Properly fitted respirators where required

- Protective skin creams when appropriate

When Exposed to a Paint Hazard

- Inhalation - Get to fresh air immediately. Oxygen or artificial respiration may be needed

- Skin Contact - Wash with soap and water after removing any contaminated clothing

- Eye Contact - Flush eyes with warm water for at least fifteen minutes and seek medical attention.

10.29   Statutory approvals, compliance and registrations

There are many statutory approvals and registrations required for project

✓   Factory Inspector – General Arrangement and Sectional Drawings of all plants and buildings inside the project need to be submitted for approval to Factory Inspector. After the approval only civil work can be started.

All Lifting tools & tackles including lifts and cranes need to be inspected by competent person appointed by Factory Inspector before putting in use.

Factory inspector visits site to review working conditions and can stop work.

✓   Municipal Corporation/ Council – Building drawings need to be submitted to Municipal Corporation/ Council in whose jurisdiction the plant falls.

✓   Atomic Energy Regulatory Board (AERB) - Nuclear Power station installation is done under supervision of AERB. In other plants, where radiography is carried out, the source pit is constructed as per AERB guidelines and operators are certified by AERB for handling radiographic source.

✓   Labour Inspector – All contractors need to take Labour license including principal contractor for engaging contract labour. Principal employer i.e. Client needs to issue Form V for this. Periodic returns need to be filed including details of migrant workers.

Labour inspector visits site periodically to ensure payment of minimum wages and maintenance of records.

✓ Electrical Inspector – All electrical installations need to be approved by electrical inspector. Single line diagram and equipment specifications need to be submitted for approval. Before charging/ commissioning the installation inspection by electrical inspector is mandatory.

✓ Boiler Inspector – All boiler installations in India need to be registered with Boiler inspector. Manufacturing drawings of boiler pressure parts and steam piping need to be approved by boiler inspector of the state where the parts are designed/ manufactured. Copies of approved drawings need to be submitted to the boiler inspector where the boiler is installed. Before lifting the pressure parts and steam piping components, boiler inspector has to inspect these parts on ground and give permission for erection. Hydraulic test of Boiler pressure parts and steam piping is witnessed and cleared by boiler inspector. Boiler inspector also witnesses floating of Boiler safety valve and issues permission to operate boiler after getting satisfied with safety valve settings.

✓ Pollution Control Board – Consent for installation and operation is required from Pollution control board. This is handled by customer.

✓ Excise Department – If any fabrication activity takes place at site then, earmarking the area and registering it is required. Fabricated material can be taken out only after making excise gate pass. Power generated from the plant or temporary DG sets also attracts levy of excise duty on units generated. Since manufacturer is claiming excise duty drawback on supplies under modVAT, scrap disposal attracts excise duty. Proper records need to be maintained.

✓ Sales Tax, VAT registration – For raising invoices on customer from site registration under Sales tax and VAT is required. Disposal of scrap is covered under these and tax should be remitted.

✓ Service Tax registration – Project site needs to be registered for service tax for raising invoices related to site work. Details of subcontractor wise payment and service tax charged by them is sought by Customs & Central Excise department. These figures are tallied with actual receipt of tax and defaulters charged with penalty.

✓ Chief Controller of Explosives (CCE) – Fuel Tank yard, Hydrogen Plant and Hydrogen cooling system for generators come under the purview of CCE. The layout and design needs to be approved by CCE.

10.30   Welding, Soldering & Brazing

After alignment the assembly can be completed by either bolting/ riveting or welding.

Welding is a fabrication process that joins materials, usually metals or thermoplastics, by causing coalescence. This is often done by melting the work pieces and adding a filler material to form a pool of molten material (the weld pool) that cools to become a strong joint, with pressure sometimes used in conjunction with heat, or by itself, to produce the weld. This is in contrast with soldering and brazing, which involve melting a lower-melting-point material between the work pieces to form a bond between them, without melting the work pieces.

Arc welding processes use a welding power supply to create and maintain an electric arc between an electrode and the base material to melt metals at the welding point. They can use either direct (DC) or alternating (AC) current, and consumable or non-consumable electrodes. The welding region is sometimes protected by some type of inert or semi-inert gas, known as a shielding gas, and filler material is sometimes used as well.

There are two basic types of weld joints

  ➤   Butt weld also called groove type joint. The joint design depends on material thickness and service condition. Edge preparation is done as per design.

  ➤   Fillet weld- Fillet welded joints such as tee; lap and corner joints are the most common connection in welded fabrication. No edge preparation is needed.

10.30.1 Power supplies

To supply the electrical power necessary for arc welding processes, a variety of different power supplies can be used. The most common welding power supplies are constant current power supplies and constant voltage power supplies. In arc welding, the length of the arc is directly related to the voltage, and the amount of heat input is related to the current. Constant current power supplies are most often used for manual welding processes such as gas tungsten arc welding and shielded metal arc welding, because they maintain a relatively constant current even as the voltage varies. This is important because in manual welding, it can be difficult to hold the electrode perfectly steady, and as a result, the arc length and thus voltage tend to fluctuate. Constant voltage power supplies hold the voltage constant and vary the current, and as a result, are most often used for automated welding processes such as gas metal arc welding, flux cored arc welding, and submerged arc welding. In these processes, arc length is kept constant, since any fluctuation in the distance between the wire and the base material is quickly rectified by a large change in current. For example, if the wire and the base material get too close, the current will rapidly increase, which in turn causes the heat to increase and the tip of the wire to melt, returning it to its original separation distance.

The type of current used plays an important role in arc welding. Consumable electrode processes such as shielded metal arc welding and gas metal arc welding generally use direct current, but the electrode can be charged either positively or negatively. In welding, the positively charged anode will have a greater heat concentration, and as a result, changing the polarity of the electrode has an impact on weld properties. If the electrode is positively charged, the base metal will be hotter, increasing weld penetration and welding speed. Alternatively, a negatively charged electrode results in more shallow welds. Non-consumable electrode processes, such as gas tungsten arc welding, can use either type of direct current, as well as alternating current. However, with direct current, because the electrode only creates the arc and does not provide filler material, a positively charged electrode causes shallow welds, while a negatively charged electrode makes deeper welds. Alternating current rapidly moves between these two, resulting in medium-penetration welds. One disadvantage of AC, the fact that the arc must be re-ignited after every zero crossing, has been addressed with the invention of special power units that produce a square wave pattern instead of the normal sine wave, making rapid zero crossings possible and minimizing the effects of the problem.

The effects of welding on the material surrounding the weld can be detrimental, depending on the materials used and the heat input of the welding process used, the HAZ can be of varying size and strength. The thermal diffusivity of the base material plays a large role—if the diffusivity is high, the material cooling rate is high and the HAZ is relatively small. Conversely, a low diffusivity leads to slower cooling and a larger HAZ. The amount of heat injected by the welding process plays an important role as well, as processes like oxyacetylene welding have an unconcentrated heat input and increase the size of the HAZ.

Many distinct factors influence the strength of welds and the material around them, including the welding method, the amount and concentration of energy input, the weldability of the base material, filler material, and flux material, the design of the joint, and the interactions between all these factors. To test the quality of a weld, either destructive or nondestructive testing methods are commonly used to verify that welds are free of defects, have acceptable levels of residual stresses and distortion, and have acceptable heat-affected zone (HAZ) properties. Types of welding defects include cracks, distortion, gas inclusions (porosity), non-metallic inclusions, lack of fusion, incomplete penetration, lamellar tearing, and undercutting. Welding codes and specifications exist to guide welders in proper welding technique and in how to judge the quality of welds. Methods such as visual inspection, radiography, ultrasonic testing, dye penetrant inspection, magnetic particle inspection, or industrial computed tomography can help with detection and analysis of certain defects.

Some of the best known welding methods include:

- Shielded Metal Arc Welding (SMAW) - also known as "stick welding", uses an electrode that has flux, the protectant for the puddle, around it. The electrode holder holds the electrode as it slowly melts away. Slag protects the weld puddle from atmospheric contamination.

- Gas Tungsten Arc Welding (GTAW) - also known as TIG (Tungsten, Inert Gas), uses a non-consumable tungsten electrode to produce the weld. The weld area is protected from atmospheric contamination by an inert shielding gas such as Argon or Helium.

- Gas Metal Arc Welding (GMAW) - commonly termed MIG (Metal, Inert Gas), uses a wire feeding gun that feeds wire at an adjustable speed and flows an argon-based shielding gas or a mix of argon and carbon dioxide (CO_2) over the weld puddle to protect it from atmospheric contamination.

- Flux-Cored Arc Welding (FCAW) - almost identical to MIG welding except it uses a special tubular wire filled with flux; it can be used with or without shielding gas, depending on the filler.

- Submerged Arc Welding (SAW) - uses an automatically fed consumable electrode and a blanket of granular fusible flux. The molten weld and the arc zone are protected from atmospheric contamination by being "submerged" under the flux blanket.

- Electroslag Welding (ESW) - a highly productive, single pass welding process for thicker materials between 1 inch (25 mm) and 12 inches (300 mm) in a vertical or close to vertical position.

Many different energy sources can be used for welding, including a gas flame, an electric arc, a laser, an electron beam, friction, and ultrasound. While often an industrial process, welding may be performed in many different environments, including in open air, under water, and in outer space.

Welding codes

The following codes are used for controlling the welding process

1 American Society of Mechanical Engineers (ASME) Codes

2 American Welding Society (AWS) Standards

3 International Organization for Standardization (ISO) Standards

These codes are used for specifications of welds, filler wires, electrodes, welder qualification and testing of welds.

Terminology

• **WPS**

- Weld Procedure Specification: Qualified instructions on how to complete the weld

• PQR

- Procedure Qualification Record (ASME) &

- Weld Procedure Approval Record: Record of the welding parameters and test results (WPAR)

• Welders Qualification Test Certificate & Welders Performance Qualification (ASME): Record of Welder test results and ranges of approval

• Essential Variable:

A parameter that when changed outside its permitted range requires requalification

• Non Essential Variable:

A parameter that when changed does not require requalification

• Supplementary Essential Variable:

An essential variable only when impact testing is required.

Format for recording WPS/ PQR as per AWS

ANNEX N AWS D1.1/D1.1M:2008

WELDING PROCEDURE SPECIFICATION (WPS) Yes

PREQUALIFIED _____ QUALIFIED BY TESTING _____

or PROCEDURE QUALIFICATION RECORDS (PQR) Yes

WELDING PROCEDURE

Identification # _____

Revision _____ Date_____ By _____

Company Name _____ Authorized by _____ Date _____

Welding Process(es) _____ Type—Manual Semiautomatic

Supporting PQR No.(s) _____ Machine Automatic

JOINT DESIGN USED POSITION

Type: Position of Groove:_____ Fillet: _____

Single Double Weld Vertical Progression: Up Down

Backing: Yes No

Backing: Backing Material: ELECTRICAL CHARACTERISTICS

Root Opening _____ Root Face Dimension _____ _____

Groove Angle: _____ Radius (J–U) _____ Transfer Mode (GMAW) Short-Circuiting

Back Gouging: Yes No Method _____ Globular Spray

Current: AC DCEP DCEN Pulsed

BASE METALS Power Source: CC CV

Material   Spec.   _____   Other _____

Type or Grade _____ Tungsten Electrode (GTAW)

Thickness: Groove _____ Fillet _____ Size: _____

Diameter (Pipe)_____ Type: _____

FILLER METALS TECHNIQUE

AWS Specification_____ Stringer or Weave Bead: _____

AWS Classification _____ Multi-pass or Single Pass (per side)_____

Number of Electrodes _____

Electrode Spacing Longitudinal _____

SHIELDING Lateral_____

Flux   _____   Gas   _____   Angle _____

Composition   _____   Contact Tube to Work Distance _____

Electrode-Flux (Class)_____   Flow Rate _____ Peening _____

_____ Gas Cup Size _____ Interpass Cleaning: _____

PREHEAT POSTWELD HEAT TREATMENT

Preheat Temp., Min. _____
Temp._____

Interpass Temp., Min. _____ Max. _____ Time _____

| Pass or Weld Layer(s) | Process | Filler Metals | | Current | | Volts | Travel Speed | Joint Details |
		Class	Diam.	Type & Polarity	Amps or Wire Feed Speed			

ANNEX N AWS D1.1/D1.1M:2008

Procedure Qualification Record (PQR) # _____

Test Results

TENSILE TEST

Specimen No.	Width	Thickness	Area	Ultimate Tensile Load, lb	Ultimate Unit Stress, psi	Character of Failure and Location

GUIDED BEND TEST

Specimen No.	Type of Bend	Result	Remarks

VISUAL INSPECTION

Appearance_____ Radiographic-ultrasonic examination

Undercut _____ RT report no.: _____ Result_____

Piping porosity _____ UT report no.: _____ Result_____

Convexity_____ FILLET WELD TEST RESULTS

Test date _____ Minimum size multiple pass Maximum size single pass

Witnessed by_____ Macroetch Macroetch

1. _____ 3. _____ 1. _____ 3. _____

2. _____ 2. _____

Other Tests   All-weld-metal tension test

Tensile strength, psi _____

Yield point/strength, psi _____

Elongation in 2 in, % _____

Laboratory test no. _____

Welder's name _____ Clock no. _____ Stamp no._____

Tests conducted by _____Laboratory

Test number _____

Per _____

We, the undersigned, certify that the statements in this record are correct and that the test welds were prepared, welded, and tested in conformance with the requirements of Clause 4 of AWS D1.1/D1.1M, (_____) Structural Welding Code—Steel. (year)

Signed _____

Manufacturer or Contractor

By _____

Title _____

Date _____

10.30.2 Welding Electrodes

In arc welding an electrode is used to conduct current through a workpiece to fuse two pieces together. Depending upon the process, the electrode is either consumable, in the case of gas metal arc welding or shielded metal arc welding, or non-consumable, such as in gas tungsten arc welding. For a direct current system the weld rod or stick may be a cathode for a filling type weld or an anode for other welding processes. For an alternating current arc welder the welding electrode would not be considered an anode or cathode.

In Gas welding, Tungsten Inert Gas (TIG) welding, MIG welding and submerged arc welding filler wire is used.

Arc welding electrodes are identified based on AWS system as

- **Welding electrode classifications**

Mild steel coated electrodes

E7018-X   E   Indicates that this is an electrode

                   70   Indicates How strong this electrode is when welded Measured in thousands of pounds per square inch.

1   Indicates in what welding positions it can in be used.

8   Indicates the coating, penetration, and current type used.

(See Classification Table below)

X   Indicates that there are more requirements.

WELDING POSITIONS

    1   Flat, Horizontal, Vertical (up), Overhead

    2   Flat, Horizontal

    4   Flat, Horizontal, Overhead, Vertical (down)

Flat Position -usually groove welds, Fillet welds only if welded like a "V"

Horizontal - Fillet welds, welds on walls (travel is from Side to side).

Vertical - welds on walls (travel is either up or down).

Overhead - Weld that needs to be done upside down.

CLASSIFICATION TABLE

Class	Electrode coating	Penetration	Current Type
Exxx0	Cellulose, Sodium	Deep	DCEP
Exxx1	Cellulose, Potassium	Deep	AC, DCEP
Exxx2	Rutile, Sodium	Medium	AC, DCEN
Exxx3	Rutile, Potassium	Light	AC,DCEP, DCEN
Exxx4	Rutile, Iron Powder	Medium	AC,DCEP, DCEN
Exxx5	Low Hydrogen, Sodium	Medium	DCEP
Exxx6	Low Hydrogen, Potassium	Medium	AC,DCEP
Exxx7	Iron Powder, Iron Oxide	Medium	AC,DCEN

| Exxx8 | Low Hydrogen, Iron Powder | Medium | AC,DCEP |
| Exxx9 | Iron Oxide, Rutile, Potassium | Medium | AC,DCEP, DCEN |

For more details one can refer to AWS.

The welding electrode identification system for stainless steel arc welding is set up as follows:

1. E indicates electrode for arc welding.

2. The first three digits indicated the American Iron and Steel type of stainless steel.

3. The last two digits indicate the current and position used.

4. The number E-308-16 by this system indicates stainless steel Institute type 308; used in all positions; with alternating or reverse polarity direct current.

- **Coatings on Welding Electrodes**

The coatings of welding electrodes for welding mild and low alloy steels may have from 6 to 12 ingredients, which include:

- cellulose to provide a gaseous shield with a reducing agent in which the gas shield surrounding the arc is produced by the disintegration of cellulose

- metal carbonates to adjust the basicity of the slag and to provide a reducing atmosphere

- titanium dioxide to help form a highly fluid, but quick-freezing slag and to provide ionization for the arc

- ferromanganese and ferrosilicon to help deoxidize the molten weld metal and to supplement the manganese content and silicon content of the deposited weld metal

- clays and gums to provide elasticity for extruding the plastic coating material and to help provide strength to the coating

- calcium fluoride to provide shielding gas to protect the arc, adjust the basicity of the slag, and provide fluidity and solubility of the metal oxides

- mineral silicates to provide slag and give strength to the electrode covering

- alloying metals including nickel, molybdenum, and chromium to provide alloy content to the deposited weld metal

- iron or manganese oxide to adjust the fluidity and properties of the slag and to help stabilize the arc

- Iron powder to increase the productivity by providing extra metal to be deposited in the weld

▪ Storage of Welding Electrodes

Electrodes must be kept dry. Moisture destroys the desirable characteristics of the coating and may cause excessive spattering and lead to porosity and cracks in the formation of the welded area. Electrodes exposed to damp air for more than two or three hours should be dried by heating in a suitable oven for two hours at 260°C. Welders are provided with portable ovens that keep the electrodes dry before use.

10.30.3 Preheating of Weld Joints

Preheating is the process applied to raise the temperature of the parent steel before welding. It is used for the following main reasons:

- To slow the cooling rate of the weld and the base material, resulting in softer weld metal and heat affected zone microstructures with a greater resistance to fabrication hydrogen cracking.

- The slower cooling rate encourages hydrogen diffusion from the weld area by extending the time period over which it is at elevated temperature (particularly the time at temperatures above approximately 100°C) at which temperatures hydrogen diffusion rates are significantly higher than at ambient temperature. The reduction in hydrogen reduces the risk of cracking.

Preheat can be applied through various means. The choice of method of applying preheat will depend on the material thickness, weldment size and the heating equipment available at the time of welding. The methods can include furnace heating for small production assemblies or, for large structural components, arrays of torches, electrical strip heaters, induction heaters or radiation heaters.

Erection Welding Schedule supplied by manufacturer gives the preheating and post weld heat treatment requirement.

Before starting any welding job, ensure availability of Welding Schedule / Erection Welding Schedule, WPS and PQR to maintain correct procedure and parameters.

10.30.4 Safety issues

Welding is a hazardous undertaking and precautions are required to avoid burns, electric shock, vision damage, inhalation of poisonous gases and fumes, and exposure to intense ultraviolet radiation.

Welding can be dangerous and unhealthy if the proper precautions are not taken. However, using new technology and proper protection greatly reduces risks of injury and death associated with welding. Since many common welding procedures involve an open electric arc or flame, the risk of burns and fire is significant; this is why it is classified as a hot work process. To prevent injury, welders wear personal protective equipment in the form of heavy leather gloves and protective long sleeve jackets to avoid exposure to extreme heat and flames. Additionally, the brightness of the weld area leads to a condition called arc eye or flash burns in which ultraviolet light causes inflammation of the cornea and can burn the retinas of the eyes. Goggles and welding helmets with dark UV-filtering face plates are worn to prevent this exposure.

10.30.5 Measurement of welding output

Any process needs to be monitored for cost effectiveness and output per person per shift/ day is one way of monitoring it. Welders are expected to deposit a minimum amount weld per shift/day.

For structural steel the welding output is measured in length of weld for a particular size of joint. This is correlated to the number of electrodes used to ensure that electrodes are fully utilised and half burnt electrodes are not wasted. Some companies enforce return of used electrode bits at the end of shift. This helps in checking proper utilisation of electrodes and at the same time proper housekeeping is also ensured.

For Tube/ Pipe welding, the concept of either equivalent joint or Inch-Dia is used to monitor welder output. Welders are required specified number of equivalent joints or Inch-Dia.

These measurements are based on weldment volume/ weight for a particular size of pipe and thickness of pipe.

In the case of equivalent joint measurement, Boiler water wall panel tube joint is taken as, 1 equivalent joint. All the other joints are converted to times equivalent joint. This helps in monitoring the progress and also prepare schedule.

The weld joint of the pipe size is converted to the equivalent inch diameter of welding. In this method the pipe is schedule 40 pipe and necessary correction should be made for other schedules of pipe.

For all pipe sizes the outside diameter (O.D.) remains relatively constant. The variations in wall thickness affects only the inside diameter (I.D.).

To distinguish different weights of pipe, it is common to use the Schedule terminology from ANSI/ASME B36.10 Welded and Seamless Wrought Steel Pipe

A schedule number indicates the approximate value of

Sch. $= 1000\ P/S$

Where

P = service pressure (psi)

S = allowable stress (psi)

The higher the schedule number is, the thicker the pipe is. Since the outside diameter of each pipe size is standardized, a particular nominal pipe size will have different inside pipe diameter depending on the schedule specified.

10.31   Heat Treatment

Heat Treatment is the controlled heating and cooling of metals to alter their physical and mechanical properties without changing the product shape. Heat treatment is sometimes done inadvertently due to manufacturing processes that either heat or cool the metal such as welding or forming.

To understand the purpose and principles of Heat Treatment, it is essential to understand the Iron Carbon Diagram. The properties of steel depend on Carbon content.

Iron Carbon Diagram

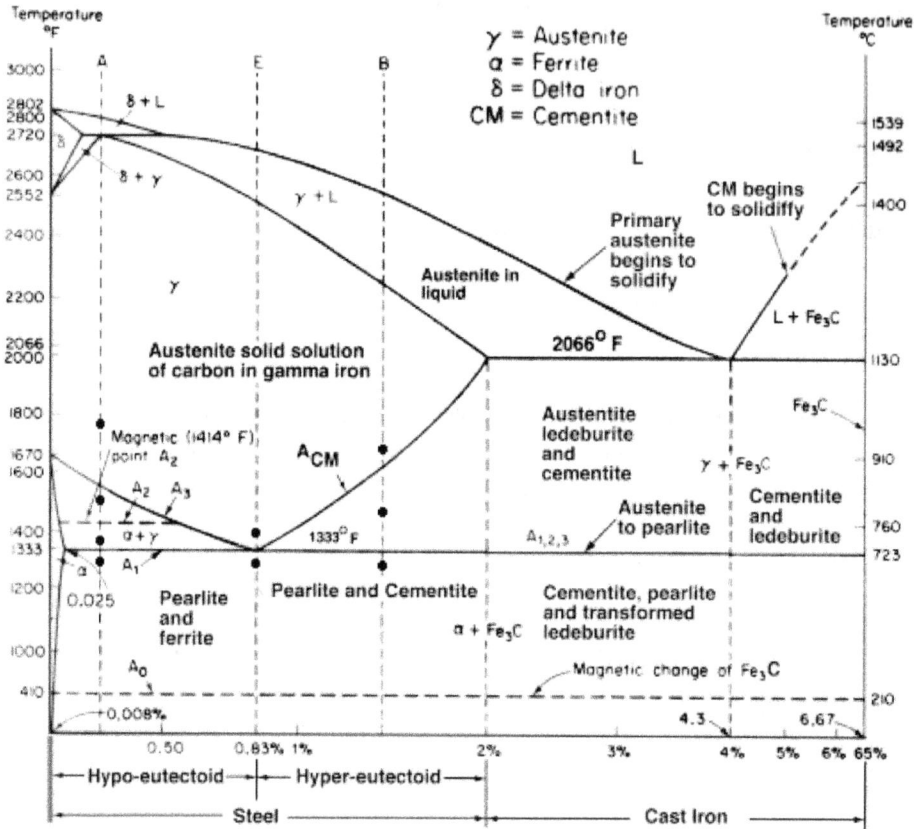

Temperature °F

Temperature °C

γ = Austenite
α = Ferrite
δ = Delta iron
CM = Cementite

L

δ + L

δ + γ

γ + L

γ

Austenite solid solution of carbon in gamma iron

Magnetic (1414° F) point A_2

A_{CM}

A_2 A_3

α + γ

1333° F

A_1

α

0.025

Pearlite and ferrite

Pearlite and Cementite

A_0

+0.008%

Austenite in liquid

Primary austenite begins to solidify

CM begins to solidify

2066° F

Austentite ledeburite and cementite

$A_{1,2,3}$

Austenite to pearlite

α + Fe₃C

Cementite, pearlite and transformed ledeburite

Magnetic change of Fe₃C

L + Fe₃C

Fe₃C

γ + Fe₃C

Cementite and ledeburite

3000
2802
2800
2720
2600
2552
2400
2200
2066
2000
1800
1670
1600
1400
1333
1200
1000
410

1539
1492
1400
1130
910
760
723
210

0.50 0.83% 1% 2% 3% 4% 5% 6% 65%
4.3 6.67

← Hypo-eutectoid → ← Hyper-eutectoid →

← Steel → ← Cast Iron →

A E B

Stress relieving is done by subjecting the parts to a temperature of about 75°C below the transformation temperature, line A1 on the diagram, which is about 727°C of steel—thus stress relieving is done at about 650°C for about one hour or till the whole part reaches the temperature. This removes more than 90% of the internal stresses. Alloy steels are stress relieved at higher temperatures. After removing from the furnace, the parts are air cooled in still air.

In Construction sites, the components are stress relieved by wrapping the portion with heating elements and thermal insulation.

- **Testing of components**

Gauges, thermocouples, display instruments and recorders need to be tested before installing to check range and calibration.

10.32   Descriptive Questions

1. Describe the various erection activities in a project.

2. Why stress relieving is required? How to determine stress relieving temperature and duration? How stress relieving is performed at erection site.

3. What are different welding processes used? What are precautions to be taken during welding?

4. Explain various types of paints used and surface preparation requirements.

5. Explain various types of fitting requirements during erection.

10.33   Multiple Choice Questions

1. Lightening Arresters are required to be installed

   i.   After completion of work

   ii.   Before starting structural erection

2. When work is done at site by deviating from drawing/ document with approval of engineering

   i.   No action needs to be taken

   ii.   As Built drawing needs to be prepared and submitted

3. Steam Blowing is carried out to

    i.   To clean Boiler internals

    ii.   To clean Turbine internals

    iii. To clean steam piping

    iv.   None of the above

4. Post weld heat treatment is required

    i.   Irrespective of thickness of material and alloy content

    ii.   When thickness of material and alloy content exceeds certain limits

    iii. None of the above

5. Oil in oil filled transformers is flushed to

    i.   To clean transformer internals

    ii.   To remove moisture

    iii. All of the above

    iv.   None of the above

State True or False

SL	Statement	True	False
01	Theodolite is used to check verticality of structures		
02	Safe working Load of slings depends on condition of sling		
03	Safety nets are required to collect debris		
04	For lifting load the selection of sling depends on equipment to be lifted		
05	Hangers need to be locked during Hydraulic test		
06	Tool box meeting is not helpful		
07	For lifting of column two cranes are used		

CONSTRUCTION – LABOUR SKILLS INVOLVED

11.1 Necessary Labour Skills

✓ Mason – Masons are required for finishing surfaces during concreting. Brick masons and stone masons are used for wall construction. Plastering is also done by masons.

✓ Carpenter- Concreting form work is prepared and fixed by carpenters. Partitions and door/ window fixing is also done by carpenters.

✓ Fitter – There are different types of fitters employed at site

✓ Civil work fitters – Prepare rebars and fix rebars in position. They also fix inserts and embedments in slabs and walls. Foundation bolt fixing prior to concreting is carried out by fitters.

✓ Structural fitters – Place and align structural steel. Bolting is also done by fitters. Cutting plan for fabrication is marked by fitters.

✓ Pressure parts fitters – Place and align pressure parts and prepare weld joints.

✓ Pipe fitters – Piping pieces are placed and aligned by these fitters. They align welded joints and release for welding. Flange fixing and bolting for bolted connections are done by these fitters.

✓ Mill Wright fitters – These fitters place, align and assemble rotating equipments. Blue matching of components is also carried out by them. These fitters align rotors and couple them. This is highest paid category in fitters.

✓ Rigger- Lifting work is carried out by riggers. They are responsible for fixing lifting tackles and signal crane operator during lifting. Specialist riggers are used for different type of lifting operations.

✓ Gas cutting – Gas cutting is done by either fitters or welders as per job requirement.

✓ Welding – Different types of welders are used in sites. These are divided in

- Structural welders – These welders do welding in civil, mechanical and electrical works. Some of them are qualified by boiler inspector to carry out attachment welding on pressure parts. These qualified welders also carry out welding on structural parts where radiography is required.

- Pressure Parts welders – These welders require permission from Boiler inspector for carrying out welding on pressure parts. Qualification in one state does not automatically qualify the welder in another state. The welders are divided into carbon steel welders, alloy steel welders and stainless steel welders. These are further divided into

    o TIG welders – These welders do TIG welding. TIG welding is carried out for root only or full joint as the drawing calls for.

    o Arc welders – These welders weld pressure parts by arc welding.

- MIG / CO2 / Aluminium welders – These are special welders. Aluminium welding is required in bus duct joints.

✓ Soldering- The quantum of soldering has come down drastically with introduction of PCBs. Nowadays no repairs are carried out on cards and cards are replaced.

✓ Brazing- Earlier generator terminals used to brazed, but with introduction of flexible connections, brazing requirement has come down.

✓ Electricians – Electrician are given license by Electrical Inspector. Different categories of certificates are issued by Electrical Inspector to electricians. Electricians should be appointed category wise in numbers required by Electricity act for handling electrical installations. For construction activity electricians are divided by voltages category as well as whether he is used for electrical work or instrumentation works.

✓ Smiths – Black smiths are not required at site. However tin smiths are required to do finer adjustments in ventilation ducts and also fabricate duct pieces to suit site conditions.

✓ Painting – Majority of painting is carried out by hand brush. Spray painting can be used in fabrication yard if site fabrication quantum is high.

✓ Drivers – Many drivers are required at project sites to drive various types of vehicles deployed.

✓ Equipment operators- There are many types of equipment deployed at site like air compressors, DG sets, vibrating needles, road rollers, pumps, concrete mixers, batching plant and compactors etc. To operate all these equipments trained operators are required.

✓ Machinist- These are required to operate lathes, saw machines, milling machines, drilling machines and grinders.

✓ Crane operators- Trained crane operators are required to operated different types of cranes.

✓ Plumbing – Plumbers are required to lay water and waste water pipe lines. They are also used to lay concreting pipe lines.

✓ Radiography – Licensed radiographic source handlers and radiographers are required at site. Along with these persons who can develop radiography films are required.

✓ Stenography & Typing – Persons with typing skills are required to type bulk correspondence. Stenographers are required for senior executives. Persons with typing skills are useful as data entry operators also.

✓ Mechanic- Trained mechanics are required for maintaining and repairing equipments.

✓ Network maintenance- Trained persons are required to set up and maintain computer networks.

✓ Winder – Winders are required for maintaining and repairing small motors.

At Hydro power stations, the stator core is built at site from stampings sent from manufacturing works. Similarly generator rotor winding also is assembled at site. For these works trained winders are required.

11.2   Descriptive Questions

1. Describe the various Labour skills required in a project.

2. Describe different types of fitting skills and their application.

11.3   Multiple Choice Questions

State True or False

SL	Statement	True	False
01	Formal certificate is a must for all trades		
02	Welder certified in one state can operate in other state without additional certification		
03	Drivers need to be certified by the type of vehicle they operate		

RISK ANALYSIS AND MANAGEMENT

Risks can come from different ways e.g. uncertainty in financial markets, threats from project failures (at any phase in design, development, production, or sustainment life-cycles), legal liabilities, credit risk, accidents, natural causes and disasters as well as deliberate attack from an adversary, or events of uncertain or unpredictable root-cause. There are two types of events i.e. negative events can be classified as risks while positive events are classified as opportunities.

Risk sources are more and more identified and located not only in infrastructural or technological assets and tangible variables, but in Human Factor Variables, Mental States and Decision Making. The interaction between Human Factor and tangible aspects of risk highlights the need to focus closely into Human Factor.

The strategies to manage threats (uncertainties with negative consequences) typically include transferring the threat to another party, avoiding the threat, reducing the negative effect or probability of the threat, or even accepting some or all of the potential or actual consequences of a particular threat.

Risk management is normally divided as

- ➤ Financial Risk Management

- ➤ Business Risk Management

- ➤ Safety Management

- ➤ Environ Management

- ➤ Security Management and

- ➤ Information Security Management

The strategies vary for each of the above. There are separate International standards for managing each of the above.

For a Project there are many risks viewed from

- ➤ Developer perspective

  - • Political Change – Political change at Centre and State or in a foreign country have impact on project. Agreements already entered may get nullified and need to be reworked having financial impact on project.

  - • Change in Government policies – Approval process and conditions may change. Tax regime may change affecting cost estimates and thereby profitability of project.

  - • RBI policy changes – Changes in RBI policy have effect on interest rates and sectoral funding. Exchange rate variation has effect on project if major equipment is imported.

  - • Delay in Financial closure of project has an effect on project schedule.

  - • Default by suppliers/ contractors – Before placing order, checking credentials of suppliers/ contractors is essential. What if analysis for different scenarios needs to be worked out to ensure adherence to schedule by suppliers/ contractors. Steps like giving out of way financial support or strengthening resources need to be taken.

- ➤ Supplier/ Contractor perspective

  - • Check financial and technical capability of Purchase before bidding for the project

- For a fixed price contract, the impact of adverse exchange rate variation, rise in input costs and Tax rate has to be absorbed in the price quoted. In a fiercely competitive market, these cannot be loaded in the price.

- Delay in financial closure of the project affects cashflow and timely payments cannot be expected. Most bidders check the status financial closure and project approvals before deciding to quote.

- In some cases the project may get shelved partly or fully during the course of contract. This affects all calculations and severe impact on a company.

- The size of contract, amount of guarantees and penalties/ liquidated damages in case of delays or foreclosure/ Termination of contract may wipe out a company if risks are not assessed properly.

- Many companies carry out a bid/ no bid exercise before taking a decision to bid

In Construction Management, Security Management is very important as the projects are in remote area and many times in politically disturbed area. There are security threats not only to project but also persons working in site both inside and outside the project premises.

Effective steps need to be taken for the safety of staff and their families. Care needs to be taken in avoiding such actions that may be construed as abating terrorist or disruptive forces.

The risk management starts with vulnerability analysis associated with threats and the steps required in mitigating the risk.

A **hazard** is a situation that poses a level of threat to life, health, property, or environment. Most hazards are dormant or potential, with only a theoretical risk of harm; however, once a hazard becomes "active", it can create an emergency situation. A hazardous situation that has come to pass is called an incident. Hazard and possibility interact together to create risk.

Identification of hazard risks is the first step in performing a risk assessment.

12.1   Common Hazards

Hazards are grouped together in following classifications

- **Safety hazards:**

These are the most common and will be present in most workplaces at one time or another. They include unsafe conditions that can cause injury, illness and death.

> Safety Hazards include:

> Spills on floors or tripping hazards, such as blocked aisles or cords running across the floor

> Working from heights, including ladders, scaffolds, roofs, or any raised work area

- **Physical hazards:**

Are factors within the environment that can harm the body without necessarily touching it.

Physical Hazards include:

> Radiation: including ionizing, non-ionizing (EMF's, microwaves, radio waves, etc.)

> High exposure to sunlight/ultraviolet rays

> Temperature extremes – hot and cold

> Constant loud noise

- **Chemical hazards:**

Are present when a worker is exposed to any chemical preparation in the workplace in any form (solid, liquid or gas). Some are safer than others, but to some workers who are more sensitive to chemicals, even common solutions can cause illness, skin irritation, or breathing problems.

Beware of:

> Liquids like cleaning products, paints, acids, solvents – ESPECIALLY if chemicals are in an unlabeled container!

> Vapours and fumes that come from welding or exposure to solvents

> Gases like acetylene, propane, carbon monoxide and helium

> Flammable materials like gasoline, solvents, and explosive chemicals.

➢ Pesticides

▪ **Ergonomic hazards:**

Occur when the type of work, body positions and working conditions put strain on your body. They are the hardest to spot since you don't always immediately notice the strain on your body or the harm that these hazards pose. Short-term exposure may result in "sore muscles" the next day or in the days following exposure, but long-term exposure can result in serious long-term illnesses.

Ergonomic Hazards include:

➢ Improperly adjusted workstations and chairs

➢ Frequent lifting

➢ Poor posture

➢ Awkward movements, especially if they are repetitive

➢ Repeating the same movements over and over

➢ Having to use too much force, especially if you have to do it frequently

➢ Vibration

▪ **Work organization hazards:**

Hazards or stressors that cause stress (short-term effects) and strain (long-term effects). These are the hazards associated with workplace issues such as workload, lack of control and/or respect, etc.

Examples of work organization hazards include:

➢ Workload demands

➢ Workplace violence

➢ Intensity and/or pace

➢ Respect (or lack of)

➢ Flexibility

➢ Control or say about things

➢ Social support/relations

➢ Sexual harassment

> ▪ **Biological hazards:**

Associated with working with animals, people, or infectious plant materials. Work in schools, day care facilities, colleges and universities, hospitals, laboratories, emergency response, nursing homes, outdoor occupations, etc. may expose you to biological hazards.

Types of things you may be exposed to include:

➢ Blood and other body fluids

➢ Fungi/ mold

➢ Bacteria and viruses

➢ Plants

➢ Insect bites, Animal and bird droppings

> ▪ **The procedure for Hazard identification and Risk assessment shall take into account**

➢ Routine and Non routine activities

➢ Activities of all persons having access to the workplace (including contractors and visitors)

➢ Human behaviour, capabilities and other human factors

➢ Identified hazards originating outside the work place capable of adversely affecting the health and safety of persons within workplace

➢ Hazards created in vicinity of workplace by work related activities under control of organisation

➢ Infrastructure, equipment and materials at the workplace, whether provided by the organisation or others

➢ Changes or proposed changes in the organisation, its activities or materials

➢ Any applicable legislation changes

➢ The design of work areas, processes etc

> - **Consider the following other conditions while identifying occupational health and safety (OHS) hazards and associated risks:**

  - ➢ DIRECT (D): Activities, which can be directly controlled by the organisation.

  - ➢ INDIRECT (I): Activities where the organisation can only have influence on its control.

  - ➢ NORMAL CONDITION (N): Risk associated with identified hazard is within tolerable limit and does not warrant any special preventive/ protective measures.

  - ➢ ABNORMAL CONDITION (AN): Risk associated with identified hazard is beyond tolerable limit and does warrant any specific preventive/ protective measures.

  - ➢ EMERGENCY CONDITION (E): Risk associated with identified hazard may lead to emergency situations and emergency response shall be invoked.

- **Risk** – The probability of an undesirable event occurring & the impact/ consequence of that event

- Risk Management –

  - ➢ Determine the probability and impact of each risk

  - ➢ Link the probability and impact of occurrence to the risk

  - ➢ Determine risk abatement action

  - ➢ Assign ownership and determine timing for each abatement action

- **Risk Management** is used in

➢ Business

➢ Quality Management

➢ Information Technology

➢ Occupational Health and Safety

➢ Environment Management

➢ Physical and Personal Security

12.2   Key Steps

- Identify the risk elements & the risk types
- Assign risk ratings to the risks: probability and consequence of risk
- Prioritize the risks – High, Medium, Low
- Identify the risk abatement plans (high & medium risks)
- Incorporate the risk abatement plan into work plans
- Track the risk score reductions & abatement actions vs plan
- Continuously update for new risks & for reduction of old risks

12.3   Ways to identify risks

- **Brainstorming of knowledgeable persons**

    This way the people share their experiences, and come out with good risk profile for any process or activity.

- **Review of lessons learnt**

    It is a good practice to write down Lessons learnt after an incident or a difficult situation. It helps in identifying underlying problems and to take corrective actions.

- **Previous experience**

    This is an important way to identify risks.

- **FMEAs**

    Failure Mode and Effect Analysis (FMEA) also called Cause and Effect diagram or Fishbone diagram is a quality management tool.

- **Previous design or producibility issues**

    Sharing knowledge of earlier faced issues, help in identifying probable risks.

- **Internet, publications and standards**

    There are lots of articles on every subject and one can get critical inputs in both identifying and mitigating risks.

12.4   Rating the risks

There are risks and there will be risks, even after taking all possible precautions. Organisations cannot afford to spend time and money on mitigating every probable risk. It is important to rate the risks and take action on significant risks.

> **▪ Pick the risk categories that apply to the risk**

- ➢ Cost
- ➢ Technology
- ➢ Specification
- ➢ Marketing
- ➢ Installation
- ➢ Legal requirement

Criteria for Risk Assessment

   A.   Over Riding Criteria

- ➢ **Domino Concern (DC):** Operations and processes which can trigger / multiply the risks those are of increasing magnitude and create potential emergency conditions.

- ➢ **Legislative Concern (LC):** Any OHS hazards / risks covered in the identification of OHS hazards and risks by existing and applicable OHS legislation and other notifications issued from time to time

- ➢ **Any Other like Chronic Effect (CE) -** Any OHS hazards / risks leads to medium and long term adverse OHS effects; Business Concern, etc

   B.   ScSe criteria

- ➢ **Sc (Scale):** Area of spread of risk i.e. Global, Regional, Local and limited to specific area of work

- ➢ **Se (Severity) :** Combination of Likelihood **(L)** and Level of Consequence **(C)**

    Where,

- **L (Likelihood): risk likelihood, probability of occurrence and**

Frequency of Exposure x Probability of Loss = Likelihood FREQUENCY OF EXPOSURE:

> **4 -** Frequent (5+ per week/job)
>
> **3 -** Occasional (1+ per week/ 3+ per job)
>
> **2 -** Rare (1+ per month/ 2+ per job)
>
> **1 - Very Rare** (1+ per year/job)

PROBABILITY OF LOSS:

> **4 -** Expected/Likely ("Happens often"; > 1 per year)
>
> **3 - Probable/** Occasional (1 occurrence in 3 years)
>
> **2 - Possible/**Rare (1 occurrence in 5 years)
>
> **1 -** Remote/Very Rare ("Once in the life of the facility or company")

- **C (level of consequence) : magnitude of harm/ damage of hazardous event**

> 5- Major and have a serious effect on the environment?
>
> 4- Significant and have a noticeable effect on the environment?
>
> 3- Moderate with a known effect on the environment? . . .
>
> 2- Limited effect on the environment?
>
> 1- Minimal with very little effect on the environment?

This methodology of risk assessment provides the classification of risks

- **Risk Assessment Total (RAT), shall be calculated for each OHS Hazards/Risks as**

$RAT = Sc + Se$, where $Se = (L x C)$

Based on **RAT**, Risk Classification shall be identified as below (the scoring legend)

> RAT 0 to 2 Trivial
>
> 3 to 6 Moderate
>
> 7 to 12 Substantial

13 to 20 Intolerable

> - **Treat any OHS Hazards/Risks having any "over-riding factor(s)" as non-tolerable OHS Risks.**
>
> - **Any OHS Hazards/Risks falling into any category other than trivial shall be treated as non-tolerable OHS Risks. ScSe score may be used to prioritize amongst the non-tolerable risks.**

12.5   Risk Assessment

Before any work commences on site risk assessments of all operations where risk is foreseeable and ensuring that appropriate control measures are established and incorporated into safe systems of work. Use these safe systems of work as the basis for the health and safety method statements. All method statements shall be developed in reasonable time to allow co-ordination of hazardous works.

The objective of risk assessment is to highlight project related hazards and to develop methods to deal with those hazards.

These assessments shall be in writing and include but not restricted to the following activities:

o Major Construction Elements

o General public and third party safety

o Location of site access/egress

o Vehicle movements on and off site

o Vehicle, Machinery and equipment hazards within the site

o Vehicle/Pedestrian segregation

o Temporary services distribution

o Siting of static plant and equipment

o Scaffolding

o Trench/Ground works

o Hazardous Chemicals

o Services Clearances

o Construction Materials

o Storage, use and disposal of substances hazardous to health

o Noise

o Working at heights

o Excavation and underground services

o Manual Handling

o Use of portable hand tools

o Emergency procedures including evacuation routes

o Fire

o Materials storage

o Site hoarding

o Contaminated ground

o Lifting new elements of structure

All risk assessments shall be reviewed and revised as necessary to accommodate any changes in methods of working, plant, equipment, material and/or site development.

12.6   Descriptive Questions

1. Describe the various types of Risks involved in a project and steps taken to mitigate them.

2. Describe the process of rating the risks and its importance in an organisation.

3. What is the difference between Risk and Hazard? Describe the various types of hazards.

12.7   Multiple Choice Questions

State True or False

SL	Statement	True	False
01	Change in government policy can affect a project		
02	Interest variation has no impact on a project		
03	Every activity needs to be analysed for associated risks		
04	If risk rating of an activity is low we need not take any action even if there is legislative concern		
05	Risk treatment means bringing potential harm to acceptable level		

ENVIRONMENTAL
MANAGEMENT REVIEW

Though the power plants are required to conduct Environmental Impact Assessment for getting clearance to install the plant, it is very important that during construction also adequate care is taken to protect the environment. By conducting the environmental review, an organisation can not only save cost but also prevent any litigation/ penalties for its operations.

13.1   Environmental Aspect

It is an 'element of an organization's activities or products or services that can interact with the environment'

Vehicle Emissions, Waste Generation, Electricity Consumption, Use of Ozone Depleting Substances, Spillage of Chemicals are examples of aspects.

13.2   Environmental Impact

It is 'any change to the environment, whether adverse or beneficial, wholly or partly resulting from an organization's environmental aspects'

Global Warming, Disposal of Waste to Landfill leading to land contamination, Ozone layer depletion and Water pollution are examples of impact.

Evaluate the significance of the Environmental Impact considering:

➢ Legal concern: Any Environmental Aspect and associated Impacts where environmental legislation and other notifications are applicable.

➢ Interested party concern: Any person/ group/ department/organization whether internal/external concerned with or affected by the environmental performance of the company

➢ Air Pollution

➢ Water Pollution

➢ Land Contamination

➢ Noise Pollution

➢ Resource Conservation / Depletion of Natural Resources

➢ Legal Compliance, Applicable legislation / Regulations

➢ Waste management

➢ Global Impacts

➢ Any other environmental concern (Pollution prevention, Interested Party concern etc.) as relevant

➢ Facilities at the work places, whether provided by the organization or others.

➢ For various environmental aspects quantify various elements for assessing environmental impact using ScSeD criteria

✓ Sc (Scale): Area of spread of environmental Impact i.e. Global, Regional, Local and specific work area

✓ Se (Severity): Level of environmental damage, magnitude of harm i.e. Catastrophic, Very serious, Serious, Major and Minor

(5-catastrophic, very dangerous, must be repaired or regenerated,

4-serious, hard to repair,

3-moderate, could be repaired,

2-minor effects that are easily remedied

1-harmless, negligible impact)

✓ **D (Duration):** Time period during which environmental impact prevails

5-More than 24 Hours

4-More than 12 hours

3-4 hours to 12 hours

2-1 hour to 4 hours

1-Less than 1 hour

✓ **Q (Quantity of aspect):** does the aspect produce:

5- A major source of emissions or use major levels of resources?

4- A high source of emissions or use high levels of resources?

3- A moderate source of emissions or moderate levels of resources?

2- A minimal source of emissions or minimal levels of resources? . . .

1- Have no source of emissions or use of resources?

Environmental Evaluation Total (EET) shall be calculated for each Environmental Aspect as given in the format using ScSeD criteria.

EET = Sc+ Se +D + Q

Significance level shall be identified as below

EET　　0 to 4 Trivial

5 to 8 Moderate

9 to 12 Substantial

13 to 16 Intolerable

Treat environmental aspects having "Overriding factor" (Legal Concern, Interested Party Concern, Business Concern etc.) as Significant Aspect.

Any environmental aspect falling into category other than Trivial/ Moderate shall be treated as Significant Aspect. **ScSeDQ** score may be used to prioritize amongst the Significant Aspect.

Review Aspect/Impact evaluation and identify the activities having significant environmental impact. Formulate necessary operation control procedure(s) to control environmental impact.

Identify HSE Management program for elimination/reduction of significant aspects indicating action plan, responsibilities and target dates for implementation.

Initial review of identification of environmental aspect and impacts shall be conducted for all activities of project. The review shall be conducted for changes/introduction of the applicable legislation/regulations.

The initial review shall be updated once in a year or earlier however the initial environmental review shall be updated if the working conditions undergo change and specific safeguards are necessitated at project during execution.

13.3   Operation Control Procedure

- For every significant Risk/ Environmental Impact identified, Control procedure needs to be developed and put in operation
- This helps in reducing the severity of accident/ incident
- Some examples of the kinds of activities that might be improved with operational controls:

> - Management/disposal of wastes
> - Approvals for using new chemicals
> - Production processes or operations
> - Storage & handling of materials and chemicals
> - Wastewater treatment
> - Building and vehicle maintenance
> - Transportation
> - Construction

13.4   Descriptive Questions

1. Describe the various types of environmental impacts involved in a project and steps taken to mitigate them.

2. Describe the process of rating the environmental impacts and its importance in an organisation.

13.5   Multiple Choice Questions

1.   For an Infrastructure project, Environmental Impact Analysis is

   i.   Mandatory

   ii.   Not Required

State True or False

SL	Statement	True	False
01	There is no problem if acid or oil spills on ground while handling		
02	Noisy operations like steam blowing can be carried out at any time		
03	Waste water can flow in to water bodies without any treatment		
04	Dust suppression measures are required to be in force during construction		
05	Welding rod bits and other scrap needs to be collected in bin and not allowed to litter		
06	Empty barrels and containers can be used for landfill		

WORK PERMIT SYSTEM

A permit-to-work system is a formal written system used to control certain types of work that are potentially hazardous. A permit-to-work is a document which specifies the work to be done and the precautions to be taken. Permits-to-work form an essential part of safe systems of work for many construction and maintenance activities. They allow work to start only after safe procedures have been defined and they provide a clear record that all foreseeable hazards have been considered.

A work permit system is a formal written system to control certain types of work identified as potentially hazardous. The terms "P.T.W.", "permit" or "work permit" refer to the form used by a company to meet its needs. These systems aim to ensure proper planning and consideration of the risks involved in a particular job, at a specific time and place, with designated precautions.

It is usually categorized in "hot" work permits and "cold" work permits. Hot works are those where there is a potential of generating fire or extreme heat, cold ones are all the others.

A permit is needed when construction or maintenance work can only be carried out if normal safeguards are dropped or when new hazards are introduced by the work. Examples are excavation, working at height, entry into vessels, hot work and pipeline breaking.

While implementing a work permit system, it is essential to do an adequacy check by answering the following

- **Is the permit-to-work system fully documented, laying down:**

- How the system works;

- The jobs it is to be used for;

- The responsibilities and training of those involved; and

- **- How to check its operation?**
- **Is there clear identification of who may authorise particular jobs (and any limits to their authority)?**
- **Is there clear identification of who is responsible for specifying the necessary precautions (e.g. isolation, emergency arrangements, etc)?**
- **Is the permit form clearly laid out?**
- **Does it avoid statements or questions which could be ambiguous or misleading?**
- **Is it designed to allow for use in unusual circumstances?**
- **Does it cover contractors?**

The following aspects should be considered with respect to Permit to Work Systems:

- **Human factors;**
- **Management of the work permit systems;**
- **Poorly skilled work force;**
- **Unconscious and conscious incompetence;**
- **Objectives of the work permit system;**

- Types of work permits required; and
- Contents of the work permits.

The following issues may contribute towards a major accident or hazard:

- Failing of the site safety management system;
- Failure to recognise a hazard before and during construction and maintenance;
- Failure to comply with the work permit system in hazardous environments; and
- Communication failures during the use of a work permit system.

Major hazards could arise from the following:

- Wrong type of work permit used;
- Wrong information about work required on the work permit;
- Failure to recognise the hazards where work is carried out (e.g. flammable substances);
- Introduction of ignition source in controlled flameproof area (e.g. welding, non-spark -proof tools, non-intrinsically safe equipment used in intrinsically safe zones);
- Terms of work permit not adhered to (e.g. failure to isolate plant and/or drain lines of hazardous substances);
- Failure to hand-over plant in safe condition on completion of work/cancelling of work permit;
- Unauthorised staff performing work permit functions;
- Poor management of the work permit system; and
- Insufficient monitoring of the work permit system.

14.1   Selection and training

- Are those who issue permits sufficiently knowledgeable concerning the hazards and precautions associated with the plant and proposed work? Do they have the imagination and experience to ask enough 'what if' questions, to enable them to identify all potential hazards?

- Do staff and contractors fully understand the importance of the permit-to-work system and are they trained in its Use?

14.2   Procedures

➤ Does the permit contain clear rules about how the job should be controlled or abandoned in case of an emergency?

➤ Does the permit have a hand-back procedure incorporating statements, that the maintenance work has finished and that the plant has been returned to production staff in a safe state?

➤ Are any time limitations included and is shift changeover dealt with?

➤ Are there clear procedures to be followed if work has to be suspended for any reason?

➤ Is there a system of cross-referencing when two or more jobs subject to permits may affect each other?

➤ Is the permit displayed at the job?

➤ Are jobs checked regularly to make sure that the relevant permit-to-work system is still relevant and working properly?

Essentials

1. Permit title

2. Job location

3. Plant identification

4. Hazard identification - including residual hazards and hazards introduced by the work

5. Precautions necessary - person(s) who carries out precautions, e.g. isolations, should sign that precautions have been taken

6. Protective equipment

7. Authorisation - signature confirming that isolations have been made and precautions taken, except where these can only be taken during the work. Date and time duration of permit

8. Extension/shift handover procedures -signatures confirming checks made that plant remains safe to be worked upon, and new acceptor/workers made fully aware of hazards/ precautions. New time expiry given

14.3 Descriptive Questions

1. Why work Permit system is required. Describe the various types of permit involved in a project and steps involved in issuing a work permit.

14.4 Multiple Choice Questions

State True or False

SL	Statement	True	False
01	Work permit need not specify time to return the same		
02	Breakers need to be drawn out before permitting work on connected system		
03	Permit can be issued by anybody		
04	Safety precautions need to be mentioned on permit		
05	There is need to monitor open work permits during shift change		

SAFE WORK PRACTICES/ OPERATION CONTROL PROCEDURES

Sl	Description
01	General safety measures
02	Safe use of safety helmet
03	House keeping
04	Electrical safety
05	Fire safety
06	Ocp for working at height
07	Working at height - safe use of scaffolds
08	Working at height - safe use of ladders
09	Handling and erection of heavy material
10	Safety in using cranes

11	Storage and handling of gas cylinders
12	Gas cutting /welding /heating
13	Manual arc welding
14	Illumination
15	Pre-commissioning and commissioning
16	Grinding operation
17	Safety precautions in gas welding/ cutting
18	Welding operation
19	Gas cutting operation
20	Material handling operation
21	Cranes & lifting eqiupments
22	Safety precautions in painting works
23	Safety in excavation
24	Safety procedures for concreting and masonary work
25	Safety procedures for working in confined spaces

15.1 General safety measures

- Ensure use of Personal Protective Equipment like safety shoes, safety helmets, safety belts, safety goggles, face shield, earplugs, hand gloves etc. as per job requirement.

- Openings in the floors must be covered.

- Install handrails on platforms and stairs as soon as these are erected.

- Temporary handrails should be provided on temporary platforms.

- Check that there are no hindrances on walkways and sufficient headroom is available.

- Warning signs should be displayed at unsafe places.

- Ensure adequate and safe means of access and exit from all work places.

- Never block accesse and exit routes to work places.

- Due importance should be given to housekeeping.

- Ensure adequate illumination of all the work places and access / exit routes.

- Arrangement of emergency lighting is also called for to ensure safety during power breakdown.

- Hazardous activity area should be cordoned off using tapes with red and white bands.

- Provide one First Aid box each for every 150 workmen.

- One vehicle should always be stationed at work site along with First Aid Box.

- Ensure adequate communication facilities between all work places and the office of Site- In charge / Emergency control room and fire station for handling emergency situation.

- Train adequate number of employees at work site in First Aid and Fire Fighting.

- Make it a habit that supervisors explain to the workers all hazard potentials and safety measures needed in the work planned for the day.

- Display appropriate posters in locally understood languages regarding safety at site.

15.2   Safe use of safety helmet

- Ensure use of Safety Helmet by all the employees working at site and the persons visiting the site, for protection against head injury.

- Safety helmets shall conform to IS: 2925

- Safety helmet should be compatible with working requirements e.g.

  - i) Helmet with chinstrap is useful for the person working in windy condition or repeated bending or looking upward.

  - ii) Helmet with head load ring is useful for the person carrying head load.

- Check the helmet regularly for any signs of damage or deterioration. Replace the helmet immediately in case positive signs of damage or deterioration are evident.

- Wear the helmet right way round. It does not give proper protection when worn back to front.

- Don't wear the helmet sloping up or down as it may significantly reduce the protection it may provide.

- Adjust the headband and the chinstrap so that there are no chances of the helmet falling down.

- Clean the helmet and its sweatband regularly and change the sweatband when required.

- Never store the helmet in direct sunlight or excessive heat. Long-term exposure can weaken the shell quickly.

- Keep stock of helmets for visitors at site. Inspect the helmet before handing it over to a visitor.

15.3   House keeping

- Unload the materials where it is required. Materials normally tend to be dumped wherever space is open or where riggers find it most convenient to unload. This lead to poor storage, poor housekeeping and damage to materials.

- Stack the raw materials like cement etc. as per standard code to avoid deterioration of materials and to ensure safety of personnel handling the materials.

- Keep the materials orderly and well stacked.

- While stacking the materials, ensure that there is no hindrance in passages for movement of people.

- Dispose of bands, binders and other packing materials as soon as supplies are unwrapped.

- Place adequate number of suitable scrap bins, biodegradable waste bins and non – biodegradable waste bins in work area.

- Remove waste /scrap, generated during erection activities regularly from working area.

- Clean the work places and passages that are slippery due to oil or water spillage. Sand or saw dust can be used for cleaning oil.

- Secure loose or light materials on open roofs and open floors against wind.

- Always keep the elevated working platforms free from unwanted tools, materials and scrap.

- Never throw materials or scrap from elevated locations to lower elevations. Take suitable measures to lower such items safely.

- The general site shall be maintained in a clean and tidy manner.

- Access ways and exits from the work place shall be kept clear of equipment, materials and garbage.

- As materials are unpacked, packaging materials, especially those of combustible nature shall be removed from the work site.

- Before entering contract, for the removal of solid wastes from site, with service contractor, verify the accreditation for such purpose and make sure that wastes are responsibly disposed of.

- The housekeeping check-list will be prepared and compliance checked periodically.

15.4 Electrical safety

- All parts of electrical installation shall be of standard construction conforming to relevant IS codes.

- ELCBs shall be provided for all the electrical systems and checked for their healthiness periodically by Safety, Electrical Engineer and Concerned contracting agency safety officer. Install Earth trip device at main distribution board.

- Earth all power cables.

- Ensure earthing of all electrical construction equipment at two locations on the body.

- Certified electricians shall carry out all electrical connections and wiring.

- Ensure that all extension boards have earth wire connected and only three-pin metal clad industrial type sockets are mounted on the board.

- Never draw power from a socket by inserting wires into the socket.

- Always use metal clad three pin plug tops for drawing power for electrical hand tools.

- Use only three core flexible wire for electrical hand tools. Connect earth wire properly to the body of the hand tool as well as to the plug top.

- Never use earth as neutral.

- Ensure that the flexible wires and power cables are properly insulated and are laid, so that the insulation does not get damaged due to activities in the vicinity.

- All cable terminations should be properly tightened.

- Cover all power cables running on surface to prevent damage due to movement of materials and vehicles.

- Install proper distribution boards at different locations to avoid haphazard connection of electrical wires.

- Frequent inspection should be carried out to identify damaged insulation, loose connections, improper fuses; lack of earthing etc. and remedial action should be initiated immediately.

- Introduce permit system for electrical safety. While men are at work on electrical lines or equipment for maintenance repair, main switch should be cut off and fuses should be removed. Warning signs should be displayed on the main switch.

- Ensure all electrical cables are free of joints by regular inspections and continuous monitoring.

- Insulation healthiness of cables is ensured by continuous checking.

- **Lock Out Tag Out** procedure shall be followed in case of repair. If anything is defective or any electrical line is under repair, a tag shall be displayed and locked in the main switch displaying the area of problem and the person who is working on the same. So any person can see that the particular work is under repair and it is locked due to that defect. The tag should contain the details of the person signing the tag, his staff number, his mobile number & name of the person. Use a standard danger tag for labelling. Destroy the tag when it is removed and a new tag will be used for other work.

- Regular joint electrical safety inspections are conducted at all the electrical systems.

- Use of correct rated fuses is always ensured.

- Check for bare conductors & Avoid joints.

- Check for broken plug & socket and replace at once.

- Check all switch boards for proper connections.

- Wear all personnel protective equipments like Safety Helmet, Safety Shoe, and Electrical Safety Gloves while working.

15.5   Fire safety

- Ensure safe access to and means of escape from all site locations.

- Construct temporary buildings like site offices and material stores, with non combustible materials.

- Keep adequate safe distance between temporary buildings. Recommended safe distance of temporary sheds from main buildings is approx 15M and 9M from each other.

- Provide free space between each section of the stored materials to allow access for fire fighting.

- Store inflammable materials separately.

- Avoid using combustible materials far from work and scaffolding.

- Remove combustible packing material from the site periodically to avoid accumulation of fire load.

- Clean site frequently at least once a week.

- Display 'No smoking' Posters in areas where inflammable materials are stored.

- Don't light any fire in or around storage areas for cooking or any other purpose.

- Inspect the work place thoroughly where welding and gas cutting operations are to be carried out and remove any inflammable materials around, before starting work.

- Ensure that welding and gas cutting sparks do not fall on combustible materials. In case it cannot be avoided, cover the materials with metal sheet.

- In case, welding or gas cutting has to be done on a vessel pipe or container, which contains or had earlier contained flammable liquids; ensure purging inerting and issue work permit before starting work.

- Ensure that there is no leakage of gases from hoses or hose connections fitted with the gas cylinders before starting gas cutting.

- In case of electric welding, check that there are no loose connections in welding cables and earth connection.

- Do not leave electric circuits on or gas cutting torches burning when not in use.

- Inspect the area for smouldering materials after gas cutting /welding. If found, extinguish them.

- Remove gas cylinders to stores when not in use.

- Place fire extinguishers in conspicuous locations as near as possible to escape routes.

- Ensure that the fire extinguishers are clearly visible and easily accessible and passage leading to them is free.

- Smoking is permitted only in designated areas. Matches/Cigarettes must be extinguished and disposed off in approved containers.

- The amount of flammable liquids/gases kept at the work area should be minimised.

- Containers of flammable liquids should be closed when not in use.

- Spills and indications of excessive flammable vapour/gas concentrations should be reported immediately.

- The necessary permits when performing hot work should be obtained.

- Materials and equipment cannot be allowed to block the access to extinguishers and fire protection hoses, hydrants, and standpipes.

15.6   OCP for WORKING AT HEIGHT

SAFETY PROCEDURES FOR WORKING AT HEIGHT

- Nobody shall be allowed to work at height (> 1.8 meter) without valid work permit and site supervisor to keep the permit while working.

- Safety Harness / 100% tie off shall be used by all workmen / supervisor, working at height.

- Sufficient access ladders, work place scaffolding, work platforms shall be arranged properly.

- Only trained man power shall be used.

- Persons working at >8 meter height are to be medically fit prior to start of work.

- The Tools / Materials taken to work place at height shall be kept properly in a holder and shall be brought down after work is over with proper checking.

- The concerned supervisor must be available at site during the period of work.

- Electrical power supply, if it concerns the place of work, shall be isolated.

- All personnel protective equipments like Safety Harness, Life Line, Fall Arrester, Safety Net, Double Shock absorbers, Full Body Harness, Helmet shall be used while working.

15.7   Safe use of scaffolds

SAFETY IN SCAFFOLD:

- Scaffold is a most commonly used access and platform for working at height.

- Scaffold is a temporary structure consisting of stairs, ladders, walkways, guard rails and toe boards.

CAUSES FOR SCAFFOLD HAZARDS

- Failure of Scaffold foundation due to
  - Soil washed away/ uncompacted soil
  - Excessive load
  - Improper bracing and tying to the structure

- Inadequate design considerations

- Inferior quality of materials used

- Improper alignment

- Use of combination of improper materials.

- Insufficient and unguarded working platforms.

- Vehicles plying at site hitting from below.

HAZARDS ASSOCIATED WITH SCAFFOLDS ARE

- Fall of persons and materials from working platform.

- People below struck by falling materials from height.

- Collapse of scaffold causing damage to property and injuries to persons.

SAFETY PRECAUTIONS

- Factor of Safety, at least 4, to be considered in design.

- Proper and sound materials are to be used.

- To be erected by experienced and trained persons under the supervision of a competent person.

- Foundation to be on firm base with adequate bracings at all levels.

- Metal scaffolds to be erected away from Over Head Transmission lines as per norms.

- Gangways / Ramps to be minimum 500 mm wide with 1:1.5 slope maximum.

- Working platforms with adequate width, guardrails and toe boards to be provided.

- Each plank must overlap the support by a minimum of 6 inches but should not extend beyond the support more than 18 inches.

- Planks must extend over the supports and overlap each other by at least 12 inches.

- If minimum clearance between the power lines and scaffolds is not possible, the power lines are to be de-energised.

- Use of electrically conductive tools or materials while working on the platforms / scaffolds must be restricted near the overhead power lines.

- Planks / platforms should be properly secured to scaffolds to prevent uplift or displacement because of high winds or other job conditions.

- Not more than one person should be allowed to stand on one plank at one time.

- Scaffolds should never be used as material hoist towers or for mounting derricks, unless the assembly is designed for the purpose.

- When necessary to prevent danger from falling objects, scaffolds should be provided with adequate screens.

- Every scaffold should, before use, be examined by a competent person to ensure more particularly:

  a) That it is in a stable condition;

  b) That the materials used in its construction are sound;

  c) That it is adequate for the purpose for which it is to be used; and

  d) That the required safeguards are in position.

Scaffolds should be inspected by a competent person and tagged clearly:

  a) Once a week.

  b) After every spell of bad weather and every prolonged interruption in the work.

  c) Scaffold parts should be inspected on each occasion before erection.

  d) No scaffold should be partly dismantled and left so that it is capable of being used, unless it continues to be safe for use.

HANDRAIL

- Height of hand rail min. 42".

- Height of the toe board min. 4".

- Working Platforms with Hand Rails shall be provided

SAFETY PRECAUTIONS

- Provide scaffolds for all works that cannot be safety performed from floor level.

- Ensure that the scaffold materials are of adequate strength for the purpose it is to be used.

- Check scaffold components for damage and defect before use.

- Avoid using combustible materials for scaffolds.

- For platforms, use planks of grade-1 quality at least 50 mm thick and 250 mm wide; free from all defects such as loose knots and splits.

- Erect scaffolds level and plumb.

- Use diagonal braces for stability of scaffolds.

- Planks used for platforms should overhang the supports by 150 – 300 mm.

- Hold the planks together by splicing at the bottom.

- Secure the platform firmly with the scaffold.

- Minimum width of platforms 2 M above floor level should be 900mm

- Provide guardrails 1.5M high, consisting of top rail and mid rail for platforms higher than 2 M from floor level.

- Provide toe guards at the platform edges to eliminate hazard of tools or other objects falling from the platform.

- Inspect the erected scaffolds regularly, if the scaffolds are to be used for a long period of time.

- Keep the platforms free from any unnecessary obstruction material, rubbish and oil spillage.

- Use safety belt properly anchored to a permanent structure while working on a platform higher than 2 M from floor level.

- Do not allow persons to work from scaffold during storm or high winds.

- Do not use braces for climbing up or down a structure or a platform. A safe and convenient means of access should be provided to all platform levels.

15.8 OCP for SAFE USE OF LADDERS

- Use portable ladders for flights up to 4 M only. Portable ladder max. 6M in length.

- Provide fixed ladders for flights above 4 M.

- Place the ladder at an angle of 75 degrees (approx) from the horizontal.

- Extend ladder at least 1 M above the top landing.

- Secure top and bottom of the ladder firmly to prevent displacement.

- Ensure that the width of the ladder is not less than 300 mm and distance between rungs is not more than 300 mm.

- Provide landings of minimum size 600 x 600 mm at intervals not more than 6 M for fixed ladders. Check the ladders daily for any defects.

- Ensure that the areas around base and top of the ladder are clear. Getting on and off the ladder is more hazardous than using it. Firm ground and strong top support is required. Use a mudsill if the ladder is to rest on soft, loose or rough soil.

- Do not use ladders of conducting material near power lines.

- Stand no higher than the fourth rung from the top for carrying out any job standing on a ladder.

- Always face the ladder while climbing up or down.

- Maintain three point contact while climbing up or down a ladder i.e. two hands and one foot or two feet and one hand on the ladder at all the times.

- Avoid climbing up or down a ladder while carrying anything in hands. Lift tools, equipment and materials with a rope.

- Ladder should be Sound in construction

- It should not be slippery

- No Make ladder shift.

- Proper ascending and descending method

- Permanent Stair case – preferable

- One Person at a time

- Ladder should be lashed at the top

- Extended 1M above the platform

- Rungs should be free from defects

- Interval between rungs max. 30cm.

- Anti-Skid hold at bottom

- Ladder shall be lashed to a fixed Structure

15.9 Handling and erection of heavy material

- Ensure proper maintenance of all the lifting machinery and tackles. Record of maintenance carried out should be available.

- Inspect daily, before start of work, all such equipment and record in a Daily Inspection Register.

- Test certificates of all the lifting machinery and tackles should be maintained. Testing shall be done periodically as recommended by the manufacturer and statutory authority.

- Maximum safe working load shall be conspicuously marked on all the lifting machinery and tackles.

- Ensure that the lifting machinery and tackles are not loaded beyond maximum safe working load.

- Before lifting with a winch ensure proper anchoring of the winch and the pulleys. The pulleys must be checked thoroughly for any defects.

- Do not tie or attach lifting devices such as chain pulley blocks, max-pull or wire rope slings to the floor grills, working platforms or guardrails.

- When load is being lifted with lifting machinery, warn the persons in the vicinity by sounding horns/ whistle.

- Do not allow anybody to stand under the load being lifted.

- Never allow any person to ride on the loads being lifted.

- Deploy only trained and authorized operators for operating the lifting machinery.

- Ensure that only skilled riggers are engaged for slinging the material to be handled.

15.10   Safety in using cranes

- Try to locate the crane on as solid and level ground as possible.

- Ensure that the earth is property compacted if the crane is required to move or work in back-filled area.

- In case there is doubt about bearing capacity of the earth, place 25 mm thick metallic plates or wooden sleepers under the crawler or tyres to distribute load.

- Deploy only trained and experienced operator with valid license to operate the cranes.

- Read the instruction manual of the crane thoroughly and keep the crane operator informed.

- Keep the machine in good operating condition especially clutches, brakes and controls.

- Ensure that boom and hoist cut off limit switches are operational.

- Before starting the crane, see that all the guards of moving parts are property fitted.

- Shut off the engine when cleaning, adjusting or lubricating the machine.

- Do not over lubricate the parts. Clean over flowing lubricants.

- Keep the crane boom length as short as possible for maximum lifting capacity and greater safety.

- Do not work with the boom too close to the vertical, because a sudden release of the load may throw the boom back over the cab.

- Depute only trained and experienced riggers for slinging the material to be lifted.

- The load on the crane shall never exceed the safe allowable working load specified by the manufactures of the crane.

- Damaged and worn lifting tools and tackles such as hooks, rings, eye bolts, chains and wire or fibre ropes and slings shall be immediately removed from service.

- Cranes and lifting equipment will only be operated and used by trained and experienced personnel. No crane or lifting device will be used on site unless it is in good condition and safe for use.

- No lifting operation will be carried out until a proper and thorough assessment has been made, having taken into account among other things the positioning of crane outriggers, slinging arrangements and banks men control.

- Cranes and lifting equipment will be clearly marked as to their lifting capacity and not used beyond that capacity. Training will be given to crane drivers and riggers stressing the necessity for a cautious approach to the lifting operations.

- The crane operators will not operate the crane until all persons concerned have been instructed as to the work to be done.

- A qualified signal man shall work with the crane operator. Standard signal must be used. Normally, the signal man must give all signals, but the operator should obey the emergency stop signal given by anyone.

- All cables should be changed when appreciable wear or corrosion is shown. Cables should be inspected daily. Sling found to be defected shall be discarded. The date on which the sling is placed in service should be stencilled on the metal connector.

- The operator must properly secure the crane and boom before or latches shall be used on crane hooks each time it is lifted.

- Load limits for all crane and "stiff legs" shall be posted in clear view of the operator. Boom angle indicators will be permanently attached to the boom to show operating radius.

- Maintenance inspection will be made monthly on cranes "stiff legs", and their wire lines.

- The lift area would be cordoned off.

- No lifting would occur when conditions exceed acceptable limits (which depend on the type of crane / lifting equipments used for that particular work).

OCP for STORAGE AND HANDLING OF GAS CYLINDERS

- While procuring gas cylinders, check that every cylinder is labelled with the name of the gas and the name and address of the person by whom the cylinder was filled.

- Check that a warning is attached to every cylinder containing permanent or liquefiable gas.

- Cylinders full or empty should not be subjected to shocks, rough usage or heating.

- Cylinders should not be exposed to sun or other sources of heat.

- Cylinders should be kept in dry place, preferably under cover and should not be allowed to remain under conditions which will cause rust or corrosion.

  i)   Do not change colour of the cylinder.

  ii)   The cylinder should not be filled with any gas other than the one it now contains.

  iii) No flammable material should be stored in the vicinity of the cylinder or in the same room in which it is kept.

  iv)   No oil or similar lubricant should be used on the valves or other fitting of the cylinder.

  v)   Please look for next date of test, which is marked on a metal ring inserted between the valve and the neck of the cylinder, and if this date is over, do not accept the cylinder.

- Use trolleys and cradles of adequate strength, as far as possible, while moving the cylinders.

- Cylinders should not be dropped. Handle the cylinders carefully and don't allow the cylinders to fall upon one another or subjected to any shock.

- Cylinders should not be lifted in loose slings and never by a lifting magnet. If cranes are used, it must have properly designed cradle with chain sling.

- Don't slide or drop the cylinders.

- Always keep LPG and other liquefiable gas cylinders in an upright position and see that they cannot be knocked over.

- Empty cylinders should be kept segregated from full cylinders.

- Valves of cylinders should remain closed and closing the valves before removing the regulator is all the more important.

- Cylinders with defective valves, with leaks should be stored in the open if the leak cannot be rectified or the gas cannot be used at once.

- Cylinders containing non-combustible gases are fitted with RH valves and combustible gases are fitted with LH valves.

- Never use oxygen as substitute for compressed air or nitrogen.

- Cylinders when not in use should always have its cap screwed in tightly for the full protection of the valve. It is extremely dangerous if the valve gets sheared off when cylinder is containing gas.

- Cylinders must be set up atleast 5 metre away from welding/ cutting torch.

- Never use cylinders as rollers, supports or for any such purpose.

- Do not place cylinders where they might become part of electric circuit.

- Avoid dragging or sliding the cylinders. Use suitable transport.

- Use only black hose for oxygen and brown hose for acetylene. Do not interchange in any case.

OCP for GAS CUTTING /WELDING /HEATING

- Use trolleys and cradles of adequate strength, as far as possible, while moving the cylinders.

- Always keep LPG and other liquefiable gas cylinders in upright position and ensure that they are not knocked over.

- Check that the valves of the gas cylinders are tightly shut when not in use.

- Do not release gas from the cylinder unless pressure regulator is fitted to its valve.

- Use gas hoses specially designed for the purpose with standard colour code.

- Use proper clamps for hose-connections. Check leakage from hose connections before starting work. Never use steel wires for clamping.

- Take care that there are no kinks in the hoses and the hoses are laid such that nobody steps on the hoses and these do not get damaged due to activities in progress in the vicinity.

- Use flame flash back arrestors to avoid back firing in flammable gas cylinders.

- Open the valve of oxygen gas first and then flammable gas for lighting the torch.

- Use friction gas lighters only for lighting the torch. Never use matches or smouldering manila ropes or rags for lighting the torch.

- Protect the gas cylinders and hoses from welding sparks or gas cutting sparks falling on them.

- Ensure that the valve key is easily accessible to close the valve immediately in case of emergency.

- Never crimp the hose for temporary shutting of gas. Always shut the supply through pressure regulators.

- Check the hoses daily for any visible damage. Discard the hoses in which gas had backfired.

- Remove the leaking cylinder of flammable gas immediately to an open space where it is least dangerous to life and property. Intimate the supplier of the cylinder.

OCP for MANUAL ARC WELDING

- Ensure proper earthing of the switch board from where power is tapped for the welding machine

- Ensure proper earthing of the welding machine by two independent earthing connections.

- Check that the terminations of the welding lead and earthing lead are not loose at the welding machine, welding holder & the job.

- There should not be more than one joint in the entire length of the welding lead. The joint should be through proper junction box or lugs. The joint should be insulated.

- Lay the welding lead in such a manner that it does not get damaged due to other activities in the vicinity.

- Provide the welder with an appropriate container for depositing welding electrode stubs, which should not be thrown indiscriminately and allowed to fall from height.

- Use of hand gloves and face shield by the welder must be ensured.

- Welding glass must conform to IS code.

- Workmen working in close association with the welder should also wear protective coloured goggles.

- Ensure that the hand gloves, welding holder and the job are not wet.

- Welder and associated workmen working at height more than 2 M must wear safety belts securely fastened to nearby structure.

- Temporary platforms/ scaffolding should be of adequate strength with handrails. Platform planks must be secured properly with the scaffolding or the structure.

- Take due care to see that welding sparks falling from height do not endanger people and property at lower elevation.

- Ensure adequate lighting and ventilation while welding in confined places. The activity should be watched through constant supervision. Switch off power supply to the welding machine during non-working period including short breaks.

OCP for ILLUMINATION

- Provide adequate illumination at all work places and their approach passages/corridors. Refer recommended values of illumination for different locations.

- Ensure that artificial lighting do not cause excessive glare of disturbing shadows.

- Provide suitable guards for the lamps where necessary to prevent danger, if lamp breaks.

- Use 24-volt supply for hand-held lamps, especially in confined spaces. Lamp guards are necessary for hand-held lamps.

- Provide emergency lighting to minimize danger in case of power failure.

Illumination - Recommended values

S. No	Location	Illumination (Lux)
	A Construction Area	
1.	Outdoor areas like store yards, entrance and exit roads.	20
2.	Platforms	50
3.	Entrances, corridors and stairs.	100
4.	General illumination of work area	150
5.	Rough work like fabrication, assembly of major items	150
6.	Medium work like assembly of small machined parts, rough measurement etc.	300
7.	Fine work like precision assembly, precision measurement etc.	700
8.	Sheet metal works	200
9.	Electrical and instruments labs	450
	B Offices	

S. No	Location	Illumination (Lux)
1	Outdoor areas like entrance and exit roads.	20
2	Entrance halls	150
3	Corridors and lift cars	70
4	Lift landing	150
5	Stairs	100
6	Office rooms, conference rooms, library reading tables.	300
7	Drawing table	450
8	Telephone exchange	200

OCP for PRE-COMMISSIONING AND COMMISSIONING

- Ensure that all construction materials, T & P, scaffolding, debris etc. are removed around the system/equipment to be commissioned.

- Ensure completion of all civil works around the system/equipment to be commissioned.

- Ensure proper approach road/gallery, platform, and stairs/ladders to the system/equipment.

- Stop welding /cutting operation in hazardous areas under commissioning.

- Ensure that all electrical installations are earthed properly before charging power supply.

- Follow the safety precautions mentioned in O & M manuals and technical procedures.

- Ensure one point control for issue of work permits.

- Ensure that all the work permits have been returned before starting the respective equipment.

- Advance announcement on Public address system if available and /or proper communication at all levels shall be ensured before switching on power to any system or equipment.

- Make available adequate fire safety measures before commissioning any system.

- Ensure that the commissioned equipment are gagged properly for identification.

- Never charge any accessible steam line, whether temporary or permanent, without thermal insulation.

- Provide earmuffs to all the personnel exposed to dangerous noise level during steam blowing operation.

- Ensure use of chemical resistant hand gloves, nose mask and safety goggles by the personnel engaged in chemical cleaning activities.

OCP for GRINDING OPERATION

- Grinding equipment like Grinding machine, grinding wheel and electrical wires shall be thoroughly checked for the damage before starting the work. Work piece shall not be forced against a wheel. Wheel speed shall be checked before changing.

- Wear all the personnel protective equipment like Safety Shoe, Leather gloves, Leg Guard, Goggles, Grinding Face Shield and Safety Helmet while working.

- Always ensure the machine operating speed does not exceed the grinding wheel speed.

- Ensure all handles, guards and safety shields are in position before starting the grinding operation.

- Don't grind on the side of the wheels, unless the wheel is specifically designed for that purpose.

- Switch off all the electrical supply when not required.

Safety Precautions in Gas welding/ cutting

- Oxyacetylene welding/cutting is not difficult, but there are a good number of subtle safety points that should be learned such as:

- More than 1/7 the capacity of the cylinder should not be used per hour. This causes the acetone inside the acetylene cylinder to come out of the cylinder and contaminate the hose and possibly the torch.

- Acetylene is dangerous above 1 atm (15 psi) pressure. It is unstable and explosively decomposes.

- Proper ventilation when welding will help to avoid large chemical exposure.

- Proper protection such as welding goggles should be worn at all times, including to protect the eyes against glare and flying sparks. Special safety eyewear must be used—both to protect the welder and to provide a clear view through the yellow-orange flare given off by the incandescing flux.

- Fuel gases that are denser than air (Propane, Propylene, MAPP, Butane, etc.), may collect in low areas if allowed to escape. To avoid an ignition hazard, special care should be taken when using these gases over areas such as basements, sinks, storm drains, etc. In addition, leaking fittings may catch fire during use and pose a risk to personnel as well as property.

- When using fuel and oxygen tanks they should be fastened securely upright to a wall or a post or a portable cart. An oxygen tank is especially dangerous for the reason that the oxygen is at a pressure of 21 MPa when full, and if the tank falls over and its valve strikes something and is knocked off, the tank will effectively become an extremely deadly flying missile propelled by the compressed oxygen, capable of even breaking through a brick wall. For this reason, never move an oxygen tank around without its valve cap screwed in place.

- On an oxyacetylene torch system there will be three types of valves, the tank valve, the regulator valve, and the torch valve. There will be a set of these three valves for each gas. The gas in the tanks or cylinders is at high pressure.

- A less obvious hazard of welding is exposure to harmful chemicals. Exposure to certain metals, metal oxides, or carbon monoxide can often lead to severe medical conditions. Damaging chemicals can be produced from the fuel, from the work-piece, or from

- a protective coating on the work-piece. By increasing ventilation around the welding environment, the welders will have much less exposure to harmful chemicals from any source.

OCP for WELDING OPERATION

- Welding cables, electrode holder, shall be thoroughly checked by Welder & Shop Incharge for any damage before starting the work.

- An unskilled labour will be available as a helper where welding is going on to watch for fire because the welder will not be aware of what is going on outside the hood.

- Wear all the personnel protective equipment like Safety Shoe, Leather gloves, Leather Apron, Leg Guard, Welding Head Shield, Full hand sleeve dress while working.

- Set the parameters for welding operation as required.

- Set the Electrode oven used for maintaining the temperature of electrode to around 150°C

- Electrode stubs shall be canned and disposed of properly and should not be thrown at site.

- Check the welding machine is properly earthed to prevent arcing.

- Check and maintain incoming electrical cable in good condition and make sure connections are tight

- Electrical equipment shall be inspected periodically and defects shall be reported to maintenance for rectification.

- Remove all inflammable materials from the location of welding to avoid fire and explosion

- Switch off all the electrical supply when not required

- Don't wrap cables around body

- Welding cable joints and connections shall be made with a connector specifically manufactured and insulated for connecting welding cables. No field splices or tape allowed on welding cables. Any welding cable with broken, cut or damaged insulation shall be removed from service immediately.

- Welding cables used for grounding shall have proper grounding clamp manufactured for that purpose.

- Hot Work Permits for Cutting and Welding will be issued by area only until the start of mechanical and electrical work after which the permits will be issued for isolated areas by everyone working in that area.

- Grounding lead clamp must be attached securely to the work piece being welded within a maximum distance of 2 meters from the welding taking place. Welding current from electrode to ground shall not pass through any motors, pumps, bearings, electrical equipment or other equipment that may be damaged by the current, welding near or around any rotating equipment whether in service or not may require special procedures to prevent damage to equipment.

- Welding cables shall not lay in water or places that may become flooded with water.

- Fire blanket shall be used to prevent sparks/hot metal from falling from height during gas cutting, welding and grinding. Fire blanket shall be used to cover equipment for protection from sparks/hot metal during gas cutting, grinding, and welding whenever necessary.

- A charged extinguisher shall be present at all times in the area where welding is being carried out.

OCP for GAS CUTTING OPERATION

- Gas cutting equipment like hoses, gas cutting torch and cylinders shall be thoroughly checked before starting the work.

- Wear all the personnel protective equipment like Safety Shoe, Leather gloves, Leg Guard, Gas Cutting Goggles, Helmet, Full Hand Sleeve dress while working.

- Ensure that the colour of LPG hose in Red and colour of Oxygen hoses in Blue.

- Cylinders, cylinder valves, hoses, regulators and gas cutting nozzles shall be kept away from oily or greasy substance.

- Always ensure that Heat / Spark / Flame are away from cylinders, articles, machinery, surfaces etc.

- No combustible substance like Paint drum, Thinner can and flaw check shall be kept nearby while gas cutting.

- Don't roll cylinders on floor; ensure use of trolley for movement of gas cylinders.

- Ensure the gas pressure for gas cutting to be around

- LPG and Oxygen cylinders shall always be kept vertical and tied off at all times.

- All cylinders shall have protective valve caps properly attached over the valves at all times when not in use.

- A charged fire extinguisher shall be available in the cutting area before any cutting is carried on.

- An unskilled labour shall be available with all cutters to keep an eye on the surrounding area for fires that might start while cutting.

- Cylinders shall not be stored in sea containers or any other such enclosed unventilated spaces.

- Hot Work Permits for Cutting and Welding will be issued by area only until the start of mechanical and electrical work after which the permits will be issued for isolated areas by everyone working in that area.

- Gas and Oxygen cylinders shall not be stored together when not in use, storage areas must maintain 6 meter min. distance of separation unless a fire wall, min 2 meter high with a fire rating of 30 min., is constructed between them.

- Cylinders shall not be transported with Gauges attached.

- Cylinders shall be protected from overhead falling objects at all times.

OCP for MATERIAL HANDLING OPERATION

- Crane, Chain slings & wire ropes shall be carefully handled and should never be overloaded and it shall not be attached to points that are not designed for lifting.

- All operations must be carried out as per coded signals given by the rigger to the crane operator. Hand signals shall be made available and the workers will be trained to use them. Crane operators will be handed with two way radio transmitters for communication.

- While operating at high angles, care shall be taken that the job does not hit the boom lattice.

- Riggers & operators shall carry a training certificate / badge so that they can be distinguished for signalling the operators.

- Always keep the hook at the centre while lifting the load. It shall not be suspended and travelled for more that 10 m and when transported, it shall be secured so that it does not swing wildly.

- The speed of mobile crane shall not go beyond 20 km / hr.

- The weight of the material can be judged before lifting and it shall be done by trained operators and riggers. The capacity of the cranes should be displayed in the machine.

- Don't make any knot in the chain sling.

- Proper equipment for handling & transporting material shall be used for that specific work for which it is designed for.

- All equipments & control switches shall be thoroughly checked before use.

- Use of wooden plank shall be ensured before placing the load on the floor / stand.

- Warning bells shall be ensured by the operator while travelling from one location to another location and on swinging any load near worker.

- Unauthorized persons shall not operate any material handling equipments.

- Do not keep the load hanging in chain slings for long time and left unattended.

- All material handling equipment shall be switched off when not required.

- Trained forklift operators shall be used for operating the forklifts. Clearance shall be analyzed first and forklifts shall be used to move around the site.

- No suspended loads shall ever be swung over the heads of any workers and anyone involved in such incidence will be immediately suspended until further action can be determined.

OCP for CRANES & LIFTING EQIUPMENTS

- Before beginning any crane operation, the supervisor and operator should complete the pre-operation checklist

- When wind velocities are above 32 km/h (20 mph), the rated load and boom lengths shall be reduced according to manufacturer specifications. All lifts above ground level, must account for wind force, i.e., side loads, down drafts, etc. as applied to the load and boom.

- Shall ensure that the relevant certification is in order prior to use. In case of a lifting appliance such as crane, the certification should confirm that it has been tested and thoroughly examined by a competent person in the preceding 12 months.

- Before any crane is used, the operator shall examine the foundation, cables, drum, brakes, boom, guards, pins, sheaves, load hook, and wire line for defects. Any defects must be repaired before the crane is used.

- Signal man and operator jointly check load charts; confirm the boom length with the chart, establish load weight, and establish the boom angle.

- There is an area surrounding every power line that is referred to as the absolute limit of approach. It is strictly forbidden to move any crane boom or load line or load into this area, unless the line has been de-energized or insulated.

Safety Precautions in Painting works

- Each product has specific safety precautions listed on the label. The following are some basic safety steps to keep in mind when using any paint.

- Always read and follow all the instructions and safety precautions on the label - DO NOT assume you already know how to use the product

  o The label will tell you what actions to take to reduce hazards and the first aid measures to use if there is a problem

- There must be plenty of fresh air where you paint- open all doors and windows to the outside (not to hallways)

  o Place a box fan in the window, blowing out to ensure air movement

  o Continue to provide fresh air after painting - ventilation should be continued for two or three days

- Follow paint can directions for the safe cleaning of brushes and other equipment

  o Latex paint usually cleans up with soap and water

  o Oil based paint require specific products as listed on the label.

  o NEVER use gasoline to clean paint brushes -gasoline is extremely flammable.

- Buy only what you need, and store or throw away the unused amount.

- If you have leftover paint, close the container tightly

- Follow directions on the paint can on how to dispose of the product

- Flammable paint must be stored in a Flammable Liquids Storage Cabinet

- Don't mix paints with other substances without approval

- Keep paints away from ignition sources and never smoke in areas where paint is used or stored

- Know where the MSDS book is kept and how to read an MSDS. Check labels of all chemicals and MSDS's for ingredients, hazard, protective procedures and PPE.

Personal Protective Equipment

- Clothing that fully covers the skin

- Gloves that resist specific paint ingredients

- Eye/face protection if recommended

- Safety glasses, goggles, hoods or face shields

- Properly fitted respirators where required

- Protective skin creams when appropriate

When Exposed to a Paint Hazard

- Inhalation - Get to fresh air immediately. Oxygen or artificial respiration may be needed

- Skin Contact - Wash with soap and water after removing any contaminated clothing

- Eye Contact - Flush eyes with warm water for at least fifteen minutes and seek medical attention.

Safe work Practice for Excavation

This document applie to all types of excavation work including bulk excavations more than 1.5 metres deep, trenches, shafts and tunnels. Excavation work generally means work involving the removal of soil or rock from a site to form an open face, hole or cavity using tools, machinery or explosives.

Excavation work may seriously affect the security or stability of any part of a structure at or adjacent to the location of the proposed excavation which can lead to structural failure or collapse. Excavation work must not commence until steps are taken to prevent the collapse or partial collapse of any potentially affected building or structure.

Any excavation that is below the level of the footing of any structure including retaining walls that could affect the stability of the structure must be assessed by a competent person and secured by a suitable ground support system which has been designed by a competent person. Suitable supports to brace the structure may also be required and should be identified by a competent person.

Precautions during excavation depend on type of soil. Ascertain type of soil before actual excavation by conducting soil investigation.

Marking all underground utilities like pipes and cables in the area to be excavated is essential. If such utility exists, excavation should be carried out in the presence of representative of the department that is responsible for the underground utility.

Before digging any trench pit, tunnel, or other excavations, decide what temporary support will be required and plan the precautions to be taken. Make sure the equipment and precautions needed (trench sheets, props, baulks etc) are available on site before work starts.

Battering the excavation sides to a safe angle of repose may also make the excavation safer. In granular soils, the angle of slope should be less than the natural angle of repose of the material being excavated. In wet ground a considerably flatter slope will be required.

Loose materials – may fall from spoil heaps into the excavation. Edge protection should include toeboards or other means, such as projecting trench sheets or box sides to protect against falling materials. Head protection should be worn.

Do not park plant and vehicles close to the sides of excavations. The extra loadings can make the sides of excavations more likely to collapse.

Edges of excavations should be protected with substantial barriers where people are liable to fall into them.

To achieve this, use:

- Guard rails and toe boards inserted into the ground immediately next to the supported excavation side; or

- fabricated guard rail assemblies that connect to the sides of the trench box

- The support system itself, e.g. using trench box extensions or trench sheets longer than the trench depth.

Ramp of proper slope should be provided for easy approach of men and equipment into the pit.

Collection pit should be made to collect water inside the pit. Drain pumps of adequate capacity should be arranged to remove water from excavation pit.

Safety in excavation:

- Soil to be tested to find out presence of any harmful gases

- Necessary arrangements are to be made to remove/ dilute such gases.

- To check for presence of underground installations, like cables, pipelines etc.

- Warning signs / Caution boards are to be displayed with proper illumination to have clear visibility during night.

- Blastings for excavation are to be carried out as per explosives handling and storing rules.

- Controlled blasting is to be carried out with cordoning off the area to prevent unauthorised entry.

- No loose Materials shall be kept very close/around excavated area.

- Proper timbering and shoring is to be ensured to prevent the loose soil sliding while the workers are working inside the pits.

- Barricading is to be provided for the excavated pits more than 1.5 metres depth.

- Movement of vehicles very close to the edge of the excavated area is to be prohibited.

- All safety equipment and signage shall be used at site like Safety Helmet, Warning Signs / Caution Board, Barricading Tape, Safety Shoes

Safety procedures for concreting and masonary work

- Stability of shuttering work shall be checked before starting concreting work.

- PPE for the all the workers should wear PPE for the work. (Like gumboots, rubber hand gloves)

- Concreting area shall be barricaded, if pouring at height / depth.

- Vibrator hoses, pumping concrete accessories shall be kept in healthy condition.

- Pipelines in concrete pumping system shall not be attached to temporary structures such as scaffolds and form work support as the forces and movement may affect their integrity.

- Safety cages / guards around moving motors / parts of concrete mixers shall be in place.

- Concrete mixers shall be provided with hoppers.

- Concrete mixers shall be inspected for their condition at start of work.

- Concrete mixers shall be maintained well so as not to generate excessive noise.

- Earthing of electrical mixers, vibrators etc shall be done and verified.

- Personal protective equipment such as gloves, safety shoe and safety helmet shall be used while dealing with concrete, and nose mask shall be used while dealing with cement.

- Cleaning of rotating drums of concrete mixers shall be done from outside. Lockout devices shall be provided where workers need to enter.

- Adequate lighting arrangement shall be ensured for carrying out concrete work during night.

- During pouring, shuttering and its supports shall be continuously watched for defects.

Safety procedures for working in confined spaces

- Confined space is an area which has a limited means of access and exit and restricted natural ventilation.

- (A sign in and sign out procedure for entry along with manned hole watch for confined space where a permit is required.)

- All work performed within a confined space must be covered by a permit to work.

- Breathing apparatus will be worn unless the area is adequately ventilated and no substances are present that will generate dangerous fumes.

- No spraying, painting or coating of substances hazardous to health is to be undertaken in any confined space, unless adequate precautions are in place to eliminate the health risk.

- No smoking, naked lights, torches, arcs, flames or other source of ignition is to be allowed within a confined space unless the atmosphere has been tested and proven safe.

- Adequate means of access and exit will be provided for all confined or enclosed spaces.

15.11 Descriptive Questions

1. Why Safe work Practices are required to be issued. Describe the various types of Operation Control Procedures used in a project.

2. What is Job Safety Analysis? How does it help in construction

3. What is a Tool Box meeting? Why it should be held before start of work? Describe the purpose and content of the Tool Box meeting.

15.12 Multiple Choice Questions

Fill in the Blanks

1. Every Lifting equipment, at site needs to be tested in presence of a _____ person as per Factory Act.

2. _____ Arrester are required for working at height.

3. During storage of rotating parts, these need to be _____ for avoiding bending of shaft.

4. While lifting a load it has to be _____ and _____.

5. Slings should be used upto their _____ working load.

6. Meggars are used for _____ testing.

7. Excavation pits need to be hard _____ to avoid fall of persons.

8. Provision of eye washers is required when process uses _____.

9. _____ charts of crane are used for deciding capacity of crane.

10. Safety posters should be in _____ language understood by workers.

State True or False

SL	Statement	True	False
01	You receive a consignment of new chain pulley block. You need to get it inspected by a Competent person before use		
02	You receive a new Hydra crane at site without registration number. It can be used immediately		
03	When people work at height, it is not necessary to cordon the area below them		
04	You can use any colour pipe for connecting pressurised gas cylinders		
05	People can work in surrounding area while radiography work is carried out		
06	Safety glasses are not required while carrying out grinding work		
07	Selection of Personnel Protection Equipment has no relation with activity being carried out		
08	Water can be used to extinguish all types of fires		
09	Noise from equipment needs to be monitored and controlled		

SITE SECURITY MANUAL

Site Security Plan for the Construction of Document No:

Project:

Contractor:

No. of Pages: _____Pages

REVISION STATUS		
Rev.	Date	Description

	Name	Designation	Date
Prepared By:			
Checked By:			
Approved By:			

Site Security Plan for the Construction of _____project
Doc. No:

Table of Contents

Sl no.	Description	Page no.
	Table of contents	
1	Purpose	
2	Scope	
3	Security facilities and infrastructure	
3.1	Perimeter fence	
3.2	All weather perimeter access road	
3.3	Watch towers	
3.4	Check post at all perimeter openings	
3.5	Transportation	
3.6	Security office, gate control posts & onsite accommodations	
3.7	Security lighting facilities	
3.8	Communications	
4	Organization and responsibilities	
5	Deployment and tours of duty	
6	Control measures	
6.1	Site access	
6.2	Personnel control	
6.3	Control of equipment, tools and construction materials	
6.4	Control of video, camera and audio equipment	
7	Security correspondence/documentation	
7.1	Reporting responsibilities	
7.1.1	Daily shift reports	
7.1.2	Perimeter patrol reports	
7.1.3	Site patrol reports	
7.1.4	Security violation reports	

16.1 Appendices

Appendix- I-Organizational Chart

A. (EPC contractor) and Site Management

B. Security Team

C. Job Descriptions for the Security Team

Appendix- II-Policies

A. Site Access Plan and Policy

B. Deployment and Tour of Duty Plan (during work hours and after work hours)

C. Personnel Access Policy Employees and Visitors

D. Vehicular Access Policy for Controlling Movement of Equipment, Tools and Materials

E. Policy for using Video, Audio Equipment

F. External Relations Policy (Dealing with the Media, Police, Fire and other outside agencies)

G. Community Relations Policy

Appendix- III-Associated Security Plans

A. Emergency Response Plan

Appendix- IV-Drawings and Sketches

A. Site Security Drawing

B. Check post Layout

1   PURPOSE

The purpose of this Security Plan is to provide an efficient and effective Security service at the _____ project site that will protect all workers and staff working at the site while implementing and enforcing the Site Security Plan set forth in this document. The two critical parts of this plan are to monitor and secure the existing perimeter and control both pedestrian and vehicle traffic entering and leaving the site through that perimeter.

(EPC contractor), in accordance with the contract, is the sole entity responsible for the security of the site and all parties working inside the site, including but not limited to, Owner, their contractors and subcontractors and all authorized visitors entering the site. All parties working inside the perimeter are expected to extend their full support and cooperation to (EPC contractor) and their Security Team to ensure that the program set forth in this Plan is implemented and followed.

Acceptance of this document does not relieve (EPC contractor) of any contractual obligations or change the contract in any way, in case of dispute or omission of information contained here, the contract shall rule.

2   SCOPE

The scope of this Plan is to provide a comprehensive list of controls (rules and regulations) that will be implemented and enforced by the Security Team and its supporting organization. The Security Team and its supporting organization are reflected in the organization charts attached Appendix I-A. Included in this Plan is a clear list of responsibilities assigned to the Security Team as well as to each individual within the organization. This Plan also defines how the security group reports through the (EPC contractor) site organization and defines the communication and documentation practices that will be followed throughout the duration of the project. The physical area covered by this Plan includes the perimeters of both the main plant and the "deep south" areas and a vital part of this Plan is patrol those perimeters and immediately report any breeches of those two perimeters.

Detailed policies, which are reviewed and revised regularly, are summarized in this Plan and detailed in the Appendices that are attached to this Plan. At any given time, the most current policy will be available through the Site Safety and Security Manager.

3.0 Security facilities and infrastructure

3.1 Perimeter Fence

Basic security of the site is provided by a 3 meter high masonry boundary wall around the perimeter of the two independent areas of the site. An access opening through the perimeter has been left on each of the four sides as shown on the Site Security Drawing in Appendix-IV-A. The east and west gates will be equipped with security check post as main entrances and the north and south gates will remain closed for now and will only be opened on request. If opened, the Security Team must have an officer on duty until such time as it can be closed

3.2 All Weather Perimeter Access Road

All weather perimeter access road is not part of the plot plan or plant layout. The perimeter access road shall be constructed early in the project to provide the Security Team with an effective means of patrolling the perimeter to be sure the perimeter is never compromised and if it is found to be compromised, then the team will notify their superior who will be responsible to get the necessary repairs completed to minimize the length of time the perimeter is compromised.

3.3 Watch Towers

A layout to be provided by EPC contractor which is delivered in the attached Site Security Drawing found in Appendix IV-A for construction of watch tower.

3.4 Check Post at all Perimeter Openings

In order to provide good control of the access through the site perimeter, EPC contractor will construct Check Post at all openings where pedestrian and vehicular traffic regularly enters and leaves the perimeter. This post will provide the Security Team with a layout that will allow them to control both pedestrian and vehicular traffic in and out of site with a minimum number of personnel while not impacting construction progress. All Check post gates will be manually operated

3.5 Transportation:

Suitable vehicles will be provided to the Security Team so they can continually monitor the perimeter as well as the work areas within the site. The vehicles provided will also be used to move the security personnel to and from their posts at shift changes and meal breaks. The vehicles shall be adequate in numbers so the security group can quickly respond to emergency situations to back up the emergency response team.

3.6 Security Office, Gate Control Posts & Onsite Accommodations:

A Security Office will be constructed to serve as a Command Post and all major access points in and out of the perimeter will have an office incorporated in the each Check post for conducting business for those who have not yet been authorized to enter the site.

The Security Team will be provided with on site accommodations so they are not adversely affected by intermingling with the villagers.

3.7 Security Lighting Facilities:

Lighting towers will be provided in areas where work is in progress & where staff/labourers are living, areas in both the Main Plant and Deep South area of the site. Light shall be sufficient in all areas to carry out routine patrols and pursue fugitives if required. All construction areas will also have additional lighting as required to safely and productively work a night shift.

3.8 Communications:

Cellular phones / Radio transmitters shall be the primary means of communication and all Security team members shall have one in their possession when on duty.

Whistles, batons, pepper spray and other such devices shall be carried by each security team member for controlling suspects or crowds.

---

4   Organization and responsibilities

Please refer to the attached organization charts in Appendix 1 for the chain of command wherein every security position is shown and a job description that defines each person's responsibilities is included in the same Appendix -1-C.

As shown in the organization charts, the Site in Charge has the ultimate responsibility for site security and he will be assisted by the Site Safety and Security Manager who will have one person, a Site Security Manager who will look after and take responsibility for the entire Security Team to see that they are properly trained, clothed and equipped to carry out their duties as set forth in this Plan. Once a night shift has started working at least a Commander shall always be on duty and accessible. The Site Safety and Security Manager and the Site Security Manager shall be authorized to liaison directly with all nearby local police and fire support agencies for assistance. On night shift the highest ranking officer onsite shall also be authorized to address a representative of an outside agency at one of the entrance gates. Please refer to Appendix II-G.

Security Team shall have a shift leader at all times who reports to the "Site Security Manager.

The Security Team has the responsibility for patrolling the overall perimeter of the work site and the Safety Supervisor who will be assisted by the team of Safety Engineers will carry out the foot patrol within the work area.

No guard shall honour any direct instruction of any (EPC contractor/ Owner) Staff Personnel. All orders and/or observations should be coursed through the Site In charge, Safety Supervisor or Safety Engineer for appropriate action.

The Security Team will be expected to work very closely with the Community Relations groups of (Owner) and (EPC contractor) to minimize any chances of being unaware of issue circulating in the villages before they become a problem. A current policy is listed in Appendix II- H.

5   Deployment and tours of duty

The Plan for deployment of the Security Team is provided in Appendix II-B and only reflects the stage of construction we are presently within. The Deployment Plan shall be updated as required in accordance with construction progress and access to the Plan will be limited to the Security Team, the Site Security Manager, the Safety and Security Manager and the top site (Owner) representative available at any given time. Policy of the deployment, the tour of duty and the timing of the deployments are referred to in Appendix II-B. Specific details of the General Orders, Code of Conduct and Code of Ethics of Security Guards should be strictly followed while on duty.

The timing for the various members of the security team will be kept to the team itself for obvious security reasons but at all times, adequate coverage will be provided 7 days a week, 365 days a year for 24 hours a day until the project is completed and accepted by (Owner).

6   Control measures

6.1 Site Access

Both pedestrian and vehicular traffic must be controlled at all times, both entering and leaving the site. Separate gates shall be used for Vehicle and labour entry. Cranes with heavy material movement will be escorted. When labour entry and exit is more, the vehicle movement will be stopped. Vehicle entry timings will be fixed and posted. Access to the site shall be strictly controlled and limited to those who have justifiable business that requires them to have to enter the site.

Offices shall be provided outside the boundary wall (or walled off inside the boundary wall) so that all administrative and sales personnel do NOT have to enter the site to carry out their business. Included in the people who will be relocated to a point outside the primary boundary wall are the community relations workers who are tasked with working with the local villages and villagers.

Pre-issued stickers will be provided for project & frequently used vehicles, a temporary gate pass will be issued for other vehicles. No vehicle shall be allowed to enter without above.

Construction traffic on the village roads must be minimized due to the risk of an accident and wherever and whenever possible roads shall be constructed inside the site to minimize the required travel on the public roads. After when the south deep gates are open the constructed road shall be used with a manned crossing for security and safety.

6.2 Personnel Control

Every person entering the site must have a badge issued by (EPC Contractor) that is visible at all times when inside the perimeter. Every person entering the site must also be wearing his PPE when entering, which shall be stored in the entry office or they will not be permitted to enter site under any circumstances. NO PERSON WILL BE ALLOWED TO ENTER THE SITE "TO GET HIS PPE". They either have, the PPE in their possession, withdraw another from the stores or have the required PPE delivered before they shall be allowed to enter the site.

(EPC contractor) shall provide a facility at the main pedestrian entrance where worker badges, visitor badges and temporary entry permits are issued. See Appendix II-C for the (EPC contractor) badging policy.

Person requesting entry as a "visitor" must first have a sponsor who works inside the plant and must either go through an on-site training session or must be accompanied 100% of the time by his sponsor. His sponsor will have to initiate the "Request for Badging" for this person and the badge to be issued can be short term or long term, depending on how often and for how long this person needs to pass through the perimeter. Refer to Appendix II-C for visitor badging requirements.

Badges issued on a temporary basis shall be immediately returned upon departure or their future access will be withdrawn until approval is again given by the Site Security Manager.

To get a temporary badge to enter the applicant must furnish his business address, nature of work inside the premises. The temporary badge should be returned at the security gate.

(EPC contractor) will control all access including the personnel of Owner or any of their visitors without any badges / escorts. The same rules that apply to EPC contractor apply to all the personnel of the aforementioned entities, no exceptions.

Emergency crews (refer to Appendix III- A) will be met at the gate by an employee to be designated by the emergency responders. The main security gate will also be contacted immediately in the case of an emergency. So should someone not show up at the gate from the site team, then security will be ready to escort the emergency teams to the scene. The Emergency Response Plan is an integral part of the site Safety and Security Plan and the most current revision is provided in Appendix III-A. All bags entering and leaving the site with a pedestrian shall be physically inspected.

6.3 Control of Equipment, Tools and Construction Materials

A current version of the Gate Pass Policy is included in Appendix II, Section D and will be updated periodically as required. Below is a quick review of the major controls to be enforced.

Equipment, tools and construction materials will be controlled going out as well as coming into site.

No materials shall be allowed to enter without first being cleared by a representative of the Stores and Quality Groups. The Construction Group will be advised through these two groups in the case of arriving materials. Inspections and certifications at the supplier's facility shall be a requirement to eliminate the chances of receiving loads that must be turned away.

Inventories will be required for all loads entering and leaving the site and will be handled by the Stores Group prior to the load entering or leaving the gate area. All equipment, tools and construction materials leaving the site must also be controlled through a GATE PASS system that requires both EPC contractor and Owner sign off. Please refer to Appendix II- D for details of the Gate Permit system.

6.4 Control of Video Camera and Audio Equipment

A current version of the Camera, Video, Audio Equipment and Mobile Phone Policies is included in Appendix II, Section E and will be updated periodically as required. Below is a quick review of the major controls to be enforced.

No cameras, video equipment or audio devices are allowed to enter the site without a written authorization of the Site in Charge or his designate. If allowed to enter, all such equipment shall also we tracked and cleared to leave site such that anything entering can be tracked to see that nothing remains on site at the end of the day. Security shall immediately report any incidents where equipment was taken inside and not removed from the site.

Badges shall be posted on the equipment brought inside by Owner. While leaving the site the badge shall be submitted in the exit gate.

7   Security correspondence/documentation:

7.1 Reporting Responsibilities

7.1.1   Daily Shift Reports shall be submitted at the end of each shift and shall designate the shift leader as well as the security officers who worked on that shift, the hours on duty, the results of inspections, emergencies, issues, problems, comments, etc.

7.1.2   Perimeter Patrol Reports

Patrolling shall be carried out a minimum of every 3 hours and a report shall be submitted daily with the Daily Shift Reports.

7.1.3   Site Patrol Reports

Patrolling shall be carried out a minimum of every 3 hours and a report shall be submitted daily with the Daily Shift Reports.

7.1.4   Security Violation Reports

For violations such as theft, illegal possession prohibited drugs or prohibited items, vandalism, trespassing, unauthorized entry, tampering of equipment and other unscrupulous acts shall be forwarded to the Site Safety and Security Manager for evaluation, proper disposition, and subsequent elevation to local law enforcement agencies for appropriate action Local Law Enforcement Agency Reports -shall be treated accordingly for purposes of legal actions, insurance claims, etc. All other security reports, as deemed necessary or as requested. (I.e. Investigation Reports, Damage Reports, Spot Reports, etc.)

All reports, wherever possible, must be completed, signed and passed onto the Site Safety and Security Manager within twenty-four (24) hours of occurrence and the incident should be informed to Owner Project manager or senior owner representative.

All correspondence/documentation as stated in this section shall be treated as "CONFIDENTIAL" and a record of the same shall be maintained and kept by the Site Safety and Security Manager Safety.

(EPC contractor) has committed to safeguard our staff and the contract workers working for (EPC contractor) to provide PPE to prevent injuries.

To enforce and implement use of Personal Protective Equipment at the site, we continuously monitor and instruct them through Tool Box Talks and daily safety Inspection. To this effect to achieve total adherence, we have prepared following action plan.

Step -1: The defaulter will be advised through safety counselling about the importance of Personal Protective Equipment at work site and will be asked to wear immediately. If not, would be sent out of the site for the day.

Step -2: If the same individual worker found second time repeating the same violation, their supervisor will be called and a warning letter will be issued in his presence as a witness.

Step –3: Third time when the same individual repeats the violation, the owner of the contractor will be called and advise him to terminate the defaulter from our service and replace.

The Site Safety and Security Manager shall have the responsibility to have the administrative support to see that all documents produced by security are properly distributed and filed for future reference with all due respect to confidentiality.

Appendix-I-Organizational Charts

    A. (EPC contractor) and Site Management

    B. Security Team

    C. Job Descriptions for the Security Team

APPENDIX-I

A   (EPC contractor) and Site Management

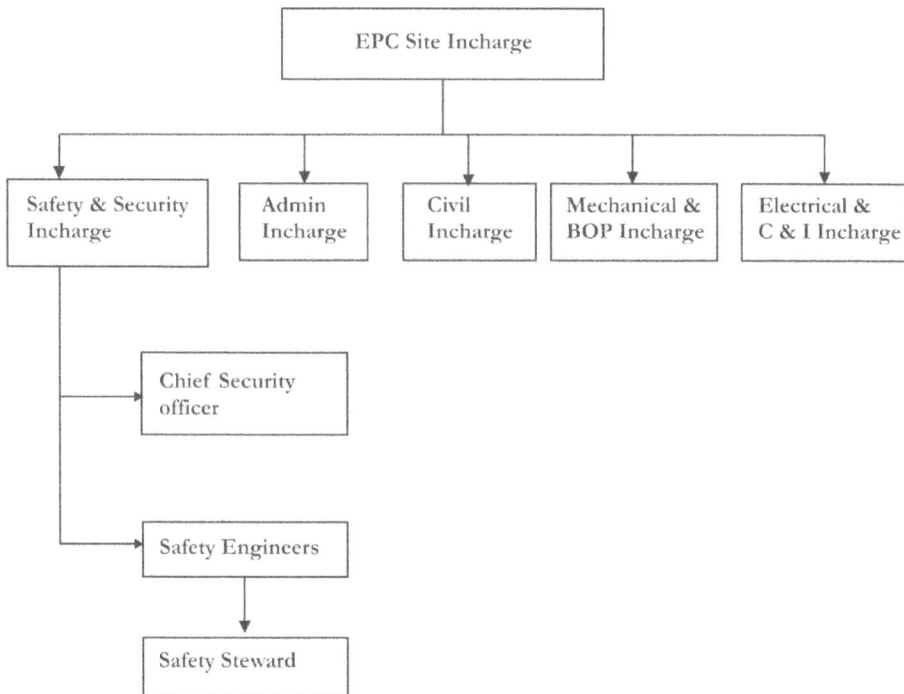

APPENDIX – I

B   **Security Team**

```
                    ┌─────────────────────────┐
                    │    EPC Site Incharge     │
                    └─────────────────────────┘
                                 │
                                 ▼
                    ┌─────────────────────────┐
                    │    Safety & Security     │
                    │        Incharge          │
                    └─────────────────────────┘
                                 │
                                 ▼
                    ┌─────────────────────────┐
                    │  Chief Security officer  │
                    └─────────────────────────┘
                                 │
        ┌────────────────┬───────┴────────┬────────────────┐
        ▼                ▼                ▼                ▼
┌──────────────┐ ┌──────────────┐ ┌──────────────┐ ┌──────────────┐
│Asst Security │ │Asst Security │ │Asst Security │ │Asst Security │
│   officer    │ │   officer    │ │   officer    │ │   officer    │
└──────────────┘ └──────────────┘ └──────────────┘ └──────────────┘
        │                │                │                │
        ▼                ▼                ▼                ▼
┌──────────────┐ ┌──────────────┐ ┌──────────────┐ ┌──────────────┐
│Security Guards│ │Security Guards│ │Security Guards│ │Security Guards│
└──────────────┘ └──────────────┘ └──────────────┘ └──────────────┘
```

APPENDIX – I

C.   ACTIVITIES AND RESPONSIBILITIES

1.   **Chief security officer**

a) He is directly responsible for control and management of the security systems in the site in accordance with the site safety plan.

b) He is responsible to submit the occurrence report of the site to the site in-charge.

c) His duty is to highlight the issues in the site to the site security manager and the site in-charge.

d) He should ensure that the security guards are well trained and he will maintain their daily attendance report.

e) He should make surprise checks to ensure proper functioning of the security system.

f) He should assume the overall responsibility to ensure that no facility is misused; ensure the material and personnel movement inside the company.

g) He should ensure whether the personal belongings of the staffs are properly accounted and permitted inside the company premises and also he will have the control of warehouse key wherever applicable.

2. Assistant Security officer

a) He is directly responsible and answerable to the CSO and he will ensure that all security personnel report for their duty on time.

b) He should ensure whether the security personnel's are well trained and give training to them under his control.

c) He should routinely visit all the duty posts to ensure alertness and correct execution of his subordinates.

d) He should ensure the control of movement of persons according to the operating procedures.

e) His responsibility is to check the vehicles entering and leaving the premises for authorization with the help of his subordinates.

f) He should have surprise checks to confirm the proper working of the security system.

g) He has to verify and maintain all the security records daily.

h) He should ensure and verify whether the company personnel working beyond the normal working hours are properly authorized and duly entered in the appropriate register.

3. Security Guard

a) Maintain the staffs and workers In/Out register and verification of personal protective equipment.

b) Maintain a well detailed vehicle entry (In/Out) register when posted at entry and exit points.

c) Maintain the raw material & pre-finished components In/Out register.

d) Monitor the work man entry and also tools equipment entry.

e) Checking the Fuel carrier vehicle entry and exit along with quantity and the fuel, oil tank measurement of DG set, Hydra and cranes.

f) When assigned to patrol duty, move around the entire site/designated area and produce patrol reports.

g) His main duty covers checking of the finished components dispatch and raw material entry with register.

h) Ensure that the vehicles are parked in their designated locations.

i) Control and monitor vehicle/crane movement inside the site.

j) Be ready to assist in fire fighting when necessary.

Appendix-II-Policies

A. Site Access Plan and Policy

B. Deployment and Tour of Duty Plan (during work hours and after work hours)

C. Personnel Access Policy Employees and Visitors

D. Vehicular Access Policy for Controlling Movement of Equipment, Tools and Materials

E. Policy for using Video, Audio Equipment

F. External Relations Policy (Dealing with the Media, Police, Fire and other outside agencies)

G. Community Relations Policy

APPENDIX –II Policies

A   Site access plan & policies

1) Purpose

The primary requirement in securing the site is the establishment of a perimeter through which access can be controlled. Herein, we will describe the perimeter as well as define what other features will be available to allow access to and from the site to be controlled. We will also spell out the policies that will be put in effect to maintain a secure access into the site.

2) Site Security Controls

a) The Security Team will be responsible to patrol and maintain the existing perimeter of both areas. Issues will be immediately brought to the attention of others for correction.

b) The second goal will be to minimize all traffic that must move in and out of the site. Those outside need to stay outside unless they have responsibilities inside and those inside need to stay inside whenever possible.

c) The third action is to then control the access as described in the following policies

3) Project Perimeter

Basic security of the site was planned to be a 3 meter high masonry boundary wall that to the start of construction and taking the custody of the plant security and safety.

The site is separated into two areas, the Main Plant area and the external area and the two are separated by a public road and therefore have their own perimeters which will have tobe patrolled and monitored. However perimeter roads are not in EPC contractor scope. The bulk of the construction activities will be in the Main Plant area except in the external area where Owner's contractor will be constructing the rail line that will be bringing coal to our plant and some housing accommodations for both (EPC contractor) and Owner.

4) Perimeter Access

As per the attached Site Security Drawing, there are presently five planned openings in the Main Plant perimeter wall with one opening still be cut through the wall. The six openings are distributed around the site with one opening on each of the sides (3nos east, 1no west, 1no north and 1no south). Presently the east gate will be primarily for pedestrian traffic and the north east gate primarily for truck traffic. The west and south gates are closed for now but will be gated for intermittent use. Long term, the south gate will likely become an opening for truck traffic to get from the main plant area to the "deep south" area in order to reduce truck traffic on the public roads around the site.

To begin with, at least the west gate and the north gates will have security check post constructed so strict security can be maintained with a minimum number of security personnel. Both pedestrian and vehicular traffic will be controlled going in and out of site as defined in SSP C-Personnel Access Control Policy and SSP D-Vehicular Access Control Policy which follow this policy.

The Deep South area, as per the drawing, will have two controlled access points, one for light duty vehicles and pedestrian traffic and another for heavy equipment traffic.

5) **Additional Security Features**

In addition to the perimeter fence, the following infrastructure will also be provided to aide security in meeting their responsibilities:

a) **All Weather Perimeter Access Road:**

Needs to construct the perimeter access road that is part of the plot plan early in the project to provide the security forces with an effective means of patrolling the perimeter to be sure the perimeter is never compromised and if it is compromised, it will be for as short of time as possible.

b) **Watch Towers:**

Will be constructed by Owner/ (EPC contractor) based on a (EPC contractor) layout which is shown in Appendix-4-1

c) **Check Posts at all Perimeter Openings:**

Check Posts will be constructed at all points where traffic regularly enters and leaves the work site.

This will provide the security team with a means to easily control both pedestrian and vehicular traffic in and out of site simultaneously, without affecting construction progress.

d) **Security Offices and Gate Control Posts:**

The Security Team will be provided with onsite accommodations as well as an independent central command centre/office. Additionally, all major access points to and from the project site will also be provided with an acceptable structure, where business other than the checking of badges, can be carried out in privacy. All security gates will be manually operated equipped with suitable equipment for checking vehicles.

Security Lighting Facilities:

Lighting towers will be provided in areas where work is in progress & living areas both the Main Plant and other areas of the site well as throughout the construction area, to effectively illuminate the job-site for Safety and Security purposes.

APPENDIX-II Policies

B   **Deployment and tour of duty plan**

1. **Purpose**

   These policies are set out to define how the facilities that are available in SSP A-Site Access Control Policy will be covered by our Security Team.

2. **Timing**

   Adequate coverage will be provided 7 days a week, 365 days a year for 24 hours a day to secure this site until the project is handed over to Owner.

3. **Tour of Duty**

   The area covered by (EPC contractor) Security Team will be limited to the interior of the boundary walls of the Main Plant and other areas.

4. **Deployment**

   The Deployment Plan is simply to patrol and maintain the perimeters, man the access gates as required to meet the policies set for in SSP C, D, E, F, and G. The deployment is also shown on the Site Security Drawing. The deployment will be updated periodically as required to accommodate the ongoing construction activities and its availability will be limited to the Security Team, the Site Security Manager, the Safety and Security Manager and the top site Owner representative available at any given time.

5. **Rules of Deployment**

   (EPC contractor) to make sure the security agency vouches for all security personnel on the Security Team engaged at this site.

   General Orders, Code of Conduct and Code of Ethics of Security Guards should be strictly followed while on duty.

   Site security officers will not respond to incidents outside the boundary walls. That responsibility rests with Owner. The site security officers will respond to a call from Owner's security if it involves an employee of (EPC contractor).

All Security Officers shall be equipped with a baton and shall carry pepper spray as a minimum to protect them.

6. **Accommodation**

All security officers will be given houses on site away from the local community to avoid being chastised because of actions taken to secure the perimeter or any other interaction with the villagers.

7. **Transportation**

Transportation will be provided on site as required to patrol the perimeters as well as the site inside the perimeter. Sufficient vehicles will also be provided for the transportation of security officers from their barracks to their work station and back as required by the day deployment plan.

Transport will also be provided to carry food to the officers on duty or to carry them to the canteen for meals. If an officer is removed from his station of duty for whatever reason, a replacement will be immediately made available.

APPENDIX – II Policies

C. **Personnel access control policy for employees and visitors**

Purpose

The purpose of this policy is to control the pedestrian traffic in and out of our site in order to permit only those who are employees or truly have business that requires that they enter the site. Reducing the number of non-employees onsite will reduce our risk of accidents that involve non-employees who have not gone through the same level of training as those employees who have that training as part of their employment.

Badging

a) The badging plan set forth in this policy statement will be implemented at the site in both areas as soon as possible and it shall be done so parallel with securing the perimeter. The badging system will enable the Security Team to be able to identify each and every person and whether he has the authorization to enter the site. The people who are allowed on site will be strictly controlled directly by (EPC contractor) through the Site Safety and Security Manager with the full cooperation of all other departments at site. Guidelines for badging shall include:

b) An application shall be submitted requesting that the person be issued a badge for entering the site. The application must designate the type of badge being requested, either a temporary badge or a long term badge.

c) Badge applications must be approved by Safety, Security, Administration as well as a senior (EPC contractor) person at site before the badge can be issued.

d) Long term badges shall be issued for a period of six months after which every employee has to have his badge renewed. Temporary badges will be issued for a shorter define period of time.

e) Only persons employed by (EPC contractor), or their designate, shall be allowed to receive a badge issued under (EPC contractor) name. Owner or their designate shall be allowed to receive a badge issued under Owner name after receiving (EPC contractor) approval.

f) A series of Visitor Badges with serial numbers will also be issued to the security group for issuance for visitors who are deemed to have real verified business inside the boundary walls.

g) The right to have a badge may be revoked at any given moment and the person holding the badge will be removed form site by security and the person shall not be allowed to re-enter the site without first having their badge reinstated.

Gate Controls

a) Pedestrian and vehicular traffic in and out of the site must be strictly controlled at all gates at all times as detailed in the Appendix-2-D to follow.

b) All bags carried by pedestrians will be checked and vehicles will be inspected before they are allowed to enter the site as well as leave the site. Vehicle inspections will be carried out on all vehicles including those owned, rented or leased by (EPC contractor) & Owner.

c) No private vehicle parking will be allowed inside the perimeter unless it is has a gate pass and the driver has a badge.

d) Every person entering the site must have an identifying badge that is pre-issued to each employee by (EPC contractor) as per the "Badging" section above. When on the site the badge must be visible at all times or the person without a visible badge will be escorted off the site. (EPC contractor) will provide the proper device to allow the badge to be properly displayed.

e) Access to the site shall be strictly controlled and limited to those who have justifiable business that requires them to physically enter the site.

APPENDIX –II Policies

D.   VEHICULAR ACCESS POLICY

Control of Equipment, Tools and Materials

Purpose

This policy is being put in place to control the access to and from site, of all vehicles that are carrying equipment, tools and materials. By enforcing this policy, incidents of theft will be drastically reduced, while at the same time, we can limit what is entering the site to what is actually approved to enter, by using representatives of the stores and quality control to release loads outside the fence before they pass through the perimeter gates.

Gate Pass

a) In order to control the flow of vehicular traffic into the site a system of Gate Passes will be put in place by (EPC contractor). Only vehicles with an approved Gate Pass will be allowed to enter or leave any perimeter gate.

b) (EPC contractor) will establish a group of personnel who will control the issuance of Gate Passes and that group will be located just outside both the (X) and (Y) gates. Pedestrian access will be available for those truck drivers & helpers wanting to leave the site. In general, trucks inside the site will be required to get their gate pass before arriving at the departure gate to avoid congestion inside the gates but plans shall be made to provide an area for a minimal number of trucks to park just inside the gate to handle those trucks that are unable to get a pass ahead of time.

c) Gate passes for vehicles coming into site as well as those going out will be signed off by the stores group as well as security.

d) Light duty vehicles belonging to the construction site will be issued permanent gate passes and the driver of that vehicle will be responsible for what he is carrying into and out of the site.

e) Commercial trucks and all other light duty vehicles will be required to have a Gate Pass. Gate Passes for light duty vehicles not belonging to the site shall also be required to have the signature of a senior site person as designate on the Authorized Signature List.

f) A staging area will be provided outside the perimeter for vehicles desiring to enter with a load. This area will prevent traffic congestion while the loads are cleared to enter.

g) Inventories will be required for all loads entering and leaving the site and will be handled by the Stores Group prior to the load entering or leaving the site.

Vehicle Controls at Perimeter Gates

a) Once any vehicle arrives at a perimeter gate to enter or leave the site, the vehicle will undergo a thorough inspection both outside and inside the vehicle.

b) All occupants will be asked to leave the vehicle and pat checking is performed before entering or leaving the site. All personal bags in the vehicle are manually checked in presence of owner whom they belong to. (Not applicable for (EPC contractor) & Owner Staffs)

c) All bags and parcels remaining in the vehicle become the responsibility of the driver and will be inspected before entering or leaving the site. The driver himself will also be pat checked before leaving or entering the site. (Not applicable for (EPC contractor) & Owner Staffs)

d) Pre-issued stickers will be provided for project & frequently used vehicles, a temporary gate pass will be issued for other vehicles. No vehicle shall be allowed to enter without the above.

Construction Traffic

Internal construction roads will be built such that on site vehicles will NOT be required to travel on the local roads any more than necessary and in particular, Owner shall provide an internal road which will carry traffic from the "Main Plant Site" to the other areas.

Construction traffic on the village roads must be minimized due to the risk of an accident and wherever and whenever possible, roads shall be constructed inside the site to minimize the required travel on the public roads.

APPENDIX-II Policies

E. Policy for using camera, video & audio equipments

Purpose

To put a consistent policy in place that will allow project management to monitor and control the flow of pictures and audio tracks such that those that do leave the project have been verified to be correct and proper.

Policy

1. No cameras, video equipment or audio devices are allowed to enter the site without a written authorization of the Site in Charge.

2. A current list of authorized camera users will be made continually available to the security staff and those EPC contractor employees and Owner employees that have been cleared to use such devices on site will have a special badge issued giving them permission to use the equipment.

3. If a temporary permit is issued to bring such equipment on site, then such equipment will be tracked by Security until it has left the site.

4. Security shall report daily, to project management, all situations where equipment was brought into site and not removed from the site as per permit conditions.

5. Owner management may authorize the use of any camera, video & audio equipment at any time.

6. Cellular phone is a mode of communication with the labourers. We can allow labourers to carry phones but if one seems to be misused by talking on phone always and taking photos of site in their mobile phones rather than working, he will be given warning for the first time and fined for the second time and dismissed from site for the third time. Winch operators will not be allowed use of mobile while operating the winch.

APPENDIX-II Policies

F.   External relations policy

Purpose

To control the flow of project information that goes to all outside agencies through a very limited number of senior EPC contractor personnel, so that project management is sure that the information being given is correct and timely.

Policy

In accordance with Policy any person trying to enter the site will be refused entry until such time as they are cleared to receive at least a Visitor's Badge and that includes representatives of the local government, the local police department, or the local fire department unless responding to an emergency situation. Members of the media also fall into this category and like the persons mentioned above, the Site Safety and Security needs to be immediately called to the gate. In his absence the Site in Charge must respond and in his absence, one of the Department Heads must respond to address the entry of these people. All staffers on site who make arrangements to have such visitor's visit the site shall provide must notify Security at least 24 hours before the arrival of that person. In the case of a pre-approved visitor, a Visitor Pass shall be issued to that person straightaway and the person who set up the meeting must travel to the gate to escort the visitors in person.

No information shall be provided to any of these outside visitors / media persons before the EPC contractor or Owner employee arranging the visit first arrives at the gate to assume control of that visitor.

APPENDIX-II Policies

G.   Community relations policy

EPC contractor will do their utmost to work closely with the Owner's Community Relations personnel to keep abreast of, and lessen, all issues directly related to impact of construction activities on the surrounding villages. EPC contractor will work closely with Owner to be proactive in meeting with the local community leaders to be sure they are heard and the issues they are tabling are addressed in some manner. Cethar stand ready to assist Owner in providing construction support within region for projects in the local community that will provide the project with a positive appearance to the villagers.

EPC contractor will do everything possible from a security standpoint to mitigate any potential impacts on the villagers caused by housing a large labour force inside the boundary wall.

EPC contractor will participate in constructing the proper facilities outside the boundary walls so the villagers can physically make contact with project personnel without having to enter the project.

EPC contractor will commit to establishing safe driving habits for all its drivers so speeds will be kept under 20kph through all villages and under 40kph on all other roads connecting the villages. Drivers not obeying these regulations will be immediately give a warning and the repeat offenders will be removed from the site.

Appendix-III-Associated Security Plans

A.   Emergency Response Plan

OBJECTIVE

The objective of this plan is to establish a standard site response and response team to deal with all emergency situations at the construction site and to make all employees aware of how they are expected to respond in an emergency situation.

INITIAL RESPONSE:

First person to recognize an emergency is expected to immediately contact the following persons:

1) Caller's first contact is Mr. _____ who can be contacted by Phone at _____.

2) Caller's second contact is Mr. _____ who can be contacted by Phone at _____.

3) Caller's third contact is Mr. _____ who can be contacted by Phone at _____.

INITIAL RESPONDER ACTIONS:

The initial responder will then contact the required emergency services as listed below:

List of emergency contacts:

- Male Nurse – Mr _____, at _____

- Ambulance - 102

- Hospital - _____

- Police - 100, Local Police Station _____

- Fire Department - 101

After the proper emergency services have been called, the initial responder is then responsible to:

See the scene of the emergency is secured either directly or by assigning the task to another responsible person. The first responder then is responsible to direct the required responsible people to meet the emergency services previously contacted and direct them straight to the scene of the emergency.

Assess the scene of the emergency looking for injures and confirming that a responsible person is taking care of all injured personnel.

After addressing all injuries, the initial responder shall assess the scene for property damage and the actions required to minimize any further property damages.

After actions have been taken to protect life and property, the initial responder shall contact the following responsible parties:

1. (EPC contractor) Site in Charge-Mr. _____ at _____

2. Mr. _____ at _____

3. Mr. _____ by phone at _____

RESPONSE TEAM:

Response Teams will be formed and trained with one response team on duty at all times that work is ongoing at site. The Response Team on duty at any given time shall be clearly posted at all entry points to the project such that, every worker entering the site to start his shift will be aware of all team members. All worker communications will be direct to the worker's immediate supervisor.

The initial site Response Team will be headed by Mr., the Site Safety Manager. As the Response Team Leader, Mr. _____ will be the first responder and will be supported by the Assistant Team Leader, Mr. _____. The other Response Team members will include M/s._____.

Mr._____, as the Assistant Response Team Leader, will assume the duties of the Response Team Leader should Mr. _____ not be available. The Response Team shall meet at least once a week to review all recent changes to the plant and procedures and to participate in a weekly training in the different aspects of properly responding. This response document is also to be updated weekly at this meeting and a current copy shall be posted all around the site in the local language as well as English, so everyone is aware of the current proper response to an emergency.

CONTROL OF SCENE:

During an emergency, Mr. _____ has the overall responsibility for controlling the site, both at the scene of the emergency and the other plant areas as per the Site Specific Safety Plan. The Line Supervisors in each area have the responsibility to control the workers in their charge to ensure that ongoing works continue and are not affected by the emergency situations that can develop.

AMBULANCE:

During the initial stages of the project, the Owner ambulance will be dedicated to the project, and (EPC contractor), based on a written joint agreement to be drafted and signed prior to the start of any works. The SKS ambulance will be the primary vehicle used to transport all injured personnel to the _____ Hospital for treatment. Prior to removing the ambulance from service at site, Owner shall inform (EPC contractor) so a substitute vehicle is available at all times. If the ambulance available at any given time does not have four wheel drive, (EPC contractor) shall assure that at least one four wheel drive vehicle is available at all times to transport the injured from the remote site to where they can be loaded into the ambulance. The initial responder shall have the responsibility to see that the _____ Hospital is advised that the ambulance will be leaving our site for the hospital with an injured person. The response team shall then advise the hospital of the condition of the injured immediately after the person leaves the scene. An emergency ambulance or an additional ambulance is available 24 hours a day by calling 102 on any mobile phone.

FIRE RESPONES:

Owner has the responsibility to provide the project fire tender and fire station. Until such time that Owner has furnished this equipment and facilities, (EPC contractor) will furnish a water tanker with a committed and operational pump dedicated to responding to a fire. The water tanker shall have a tractor connect to it at all times and may be used for other construction activities as long as it is clear to everyone that in case of a fire this tanker immediately responds.

During off hours the water tanker shall be full and left connected to the tractor for any required responses during non working hours. The tractor driver shall be communicated through phone / radio and his number will be posted for emergency purposes.

Appendix-IV-Drawings and Sketches

    A.   Site Security Drawing

    B.   Check post Layout

Appendix-IV-Drawings and Sketches

A. Site Security Drawing

Appendix-IV-Drawings and Sketches

    B.   Check post Layout

16.2 Descriptive Questions

1. Why Project specific Site Security manual is required to be issued. Describe the various types of security threats in a project.

16.3 Multiple Choice Questions

State True or False

SL	Statement	True	False
01	Gate control is an important function of Security		
02	VIP vehicles need not be checked at gate		
03	Security guards can manhandle a culprit		
04	Security guards are accommodated in Labour colony		
05	Entrance for vehicles and personnel can be same		
06	Any agency can be engaged for providing security services		

SITE SAFETY MANUAL

COMPANY'S OHSE POLICY

सुरक्षा की शपथ

मैं आज मेरे काम को सुरक्षित रूप से करूँगा और
मेरे साथ काम करनेवाले भाईयौं को भी कोई
मुसिबत मे न आने दुंगा और कार्यदक्षा से
उत्पादन बढ़ाने का शपत लेता हूँ ।

SAFETY PLEDGE

Today, I will do my job safely without
creating any danger to me as well as to
my co-workers and take pledge in doing my
job efficiently to improve the production.

1. INTRODUCTION & SCOPE

1.1 In the (Name of the project) project (name of EPC contractor) is committed to carrying out safe works in all activities. To achieve this, the project must ensure the health, safety and environment of those involved in and around the works, including the general public, and provide adequate protection for the environment.

1.2 This Project plan outlines the Health and Safety Management System to be implemented by project HSE team for all Works (including works of suppliers and subcontractors as appropriate) on the project. Construction team will implement the HSE program with the HSE site team overseeing that construction is proper and implemented what is in this manual.

1.3 New areas of safe procedure that are not covered in the CSSM, will be added to this manual by EPC contractor with Client approval. This document is subject to changes with owner approval.

2 STRUCTURE AND RESPONSIBILITES:

- The safety organization chart is given in Appendix. This chart would be reviewed and updated when there is any change of the personnel in HSE activities.

- This Project manual meets or exceeds the Corporate Safety Manual in all aspects. (EPC contractor has an HSE management system and it shall be made clear to all levels in the organization and commitments from the relevant personnel, who are responsible and accountable for Implementing / Monitoring HSE performance and activities being carried out as planned and within the scope of their responsibilities.) The roles and Responsibilities to be played by all key personnel as listed are to be followed.

2.1 PROJECT MANAGER / SITE INCHARGE:

He is responsible for

- The overall management and control of the project persons to achieve site safety performance and follow HSE activities in an efficient manner.

- Execution planning and control of all activities and reporting the progress status.

- Site safety plan implementation through involvement of construction.

- Allocation of resources and fix priority for implementation of safety plan & oversees the procurement of safety items but the control of the inventory will be controlled by SSM.

- Safety performance expectations and safety promotional activities are to be communicated to the entire site team including contractors.

- Monitor customer feedback, complaints and ensure action has been taken in a timely manner.

- Managing the requirements, interface with Client, Contractors, Sub-contractors and Local Authorities

2.2 SITE SAFETY MANAGER (SSM):

He is responsible for

- Ensure all accident and dangerous occurrences are reported immediately.

- Direct and monitor the Corrective / Preventive action and follow up activities.

- Recognize outstanding safety performance in the group to increase commitment and participation.

- Identify define, implement and monitor the entire site safety requirement through construction.

- Monitor safe work practices for continuous improvement.

2.3 SITE SAFETY ENGINEER:

He is responsible for

- Coordinating and assisting Project Manager in working out the program in accordance with Safety Plan.

- To review the program periodically and discuss the result achieved.

- Conduct inspection periodically, by using safety check list.

- Initiating Spot corrective action, for all unsafe conditions and unsafe acts at the site.

- Take suitable action to eliminate findings of unsafe work at the site.

- Organizing tool box talks / Safety Pledge on daily basis.

- Organize safety induction training to all at site

- Organize safety committee meetings.

- Organize safety committee inspection.

- Maintain accident records.

  1. First aid cases

  2. Incidents.

  3. Accidents.

- Investigate all accidents and recommend remedial measures to avoid such accident. Implementation of remedial measures in co-ordination with site incharge.

- Discuss all accidents in the safety committee meeting.

- Display safety slogans / posters in all conspicuous locations of the site.

- Organize safety promotional activities including Safety Day Celebrations.

- Plan the requirements of first aid, fire extinguishers and PPE and implement all in site to ensure safety.

- Organize safety training to sub-contractors, contractors, supervisors and officials at all levels.

- Prepare monthly safety report, weekly safety report and submit to Site In charge and the Client.

2.4 SITE ENGINEER / SUPERVISIOR:

They are fully responsible for safety of their workmen and others working in their control.

- Assist in the implementation of the "Site Safety Plan".

- To take immediate action for correcting the unsafe conditions or actions in the site.

- Implement all safety rules and regulations as per Local Law and as advised by Site Safety Engineer.

- Participate in safety meetings.

- Participate in Job safety Analysis and Safety assessments as required by the HSE department at site.

- To intimate all accidents, incidents to the site Incharge, SSM & Site Safety Engineer.

- Arrange First Aid / Treatment to injured person immediately.

- Assist the safety engineer in conducting accident investigation and preparing required reports.

- Enforce safety rules and take action to ensure compliance.

- Evaluate the safety performance of the assigned employees and report the finding to the Site In-charge for striving to improve the work practices.

- Conduct drug and alcohol fitness for duty.

- Attend the safety trainings, as an example to others working in their group.

- He shall obey all safety norms as an example to others working in their group.

- To ensure proper housekeeping in their respective work area.

- Ensure usage of recommended PPE by all in the site.

2.5. CONTRACTORS AND SUB-CONTRACTORS:

They are responsible for following all safety precautions at the site by their workers.

- Actively participating in the Accident prevention programs arranged by Safety Engineer.

- Knowing all Hazards in their work place & educate their workers about the same.

- Attending safety meeting organized by the site safety engineer.

- Ensuring new employees attending safety induction training and specialized training.

- Actively participating in the safety committee meeting.

- Ensuring all necessary PPE's are provided and used by all their workmen.

- Conducting daily and weekly assessment for compliance with safety requirement.

- Educate all workers in keeping their workplace neat & tidy.

- Arrange for taking Safety Pledge daily.

- Client will recognize all Contractors and Subcontractors as EPC contractor only. Hence EPC contractor is responsible for all Safety and Environmental requirements by its subcontractors.

- If the labour does not follow the rules, he will be given warning for the first time. If he does the same mistake again one day attendance will be lost by him. If the same labour does the same thing, he will be terminated from the site.

2.6. WORKERS:

- Working in a safe manner at all times.

- To understand the job and to do the same in a safe manner.

- Making safety suggestions and reporting unsafe conditions to the Site Safety Engineer.

- Maintain safe work practices and accident free atmosphere.

- Reporting substandard practices, conditions or behaviours to their supervisor.

- Promptly reporting injuries or incidents to their supervisors and Safety Engineer, including Near Miss Incidents.

- To ensure correct usage of safety equipment including PPEs.

- Participate in all Safety Promotional activities.

- To give feedback about the Personal Protective Equipment.

2.7 RULES FOR VISITORS:

Visitors entering the site shall record the name in the register at the security gate, the purpose of their visit and the person to whom they want to meet shall be entered in the register.

- Time of entry and exit will be entered in the register.

- Visitors will ensure wearing of helmet when they are needed to make an entry in the helmet - zone area.

- Visitors vehicle if driven on site shall be escorted or have site assigned escort inside vehicle at all times.

- Visitors with valid permit shall be allowed unless the escort confirms the absolute requirement that the Visitor must be allowed to enter the site.

- All Visitors who have not completed the site specific safety induction / orientation and has been issued site badges shall be escorted at all times while on site by the person who has invited the Visitor, the person shall be responsible for the Visitor and Visitor's action while on site.

- Visitors found not following any safety rule shall be immediately removed from site and his rights for further access must be withdrawn unless they undergo further training and signoff, stating that their entry is only on probation and any further infractions will result in immediate dismissal.

- Visitors shall follow all instructions given by the Site Incharge.

- Visitors shall ensure they visit only the permitted area & they shall be watchful throughout the site visit.

- Visitors will not be allowed on site without a badge and an escort unless they have undergone training.

- Visitors shall ensure their vehicle parked only in the vehicle parking area, without affecting other vehicle movement.

- Drive their vehicle, within speed limit as per site rule.

- Unauthorized way of working in the site shall be totally eliminated.

- All Visitors shall comply with all site safety regulations, but not limited to wearing of all required PPE in certain specified areas.

2.8. RESPONSIBILITY:

"Health, Safety and Environment is everyone's responsibility" The SSM oversees the training programs, monitors the execution of the safety plan and reports on failures and areas that need improvement.

3.   STOP WORK ORDERS:

    3.1 The Project in Charge or Site Safety & Security Manager is authorized to issue a 'Stop Work Order' if:-

    1.   site conditions or operations create a significant risk to health, safety, or the environment

2   site conditions or operations create a significant risk of examination

3   safety procedures/instructions are not followed satisfactorily and

4   when an accident / incident occurs.

3.2 'Stop Work Orders' are initially issued verbally to the senior person in charge of the relevant works, area, materials, equipment and etc. The verbal 'Stop Works Order' shall be followed up with a confirmation memo within one hour from the verbal instruction. Powers shall be expanded to the Owner's Agent in the event of potential risk situation and anyone in the case of a life and death situation. All "Stop Work" orders, despite where they come from, shall be immediately investigated as to the "reasonableness" of the order and records shall be kept to spot regular violators and multiple " Stop Orders" for the same reason or involving the same group of people or individuals.

4.   SAFETY AWARENESS AND TRAINING:

The training needs are identified by Site In charge and Site Safety Engineer and shall provide appropriate training for all personnel working in the site.

- The company HSE policy will be explained to all site personnel as part of the Site Induction.

- The HSE policy will be displayed at a well-known location on site, if practical, e.g. site office, notice board, container, large toolkits.

- Personnel performing specific assigned task shall be competent on the basis of appropriate education and training.

- The contract workers will be given training for acquiring knowledge and skills to perform work in a safe and environmentally responsible manner.

- The training needs shall be identified based on the requirement of OHSE policy.

- Emergency Preparedness & Response requirements and potential consequence of deviation from safe operating procedure shall be educated.

Safety awareness / Training program shall include the following.

1. Safety induction training for new employees.

2. Daily Tool box talk / Safety Pledge.

3. Safety training for supervisors and engineers.

4. Display & propagation of OHSE policy at site.

5. Display of Accident statistics.

6. Display of safety posters.

7. Specialized trade wise safety training.

8. Fire fighting and fire control training.

9. First Aid Training.

10. Weekly joint meetings in each of the work areas and monthly meetings for the entire project

5.   SITE SAFETY COMMITTEE:

- The site shall have a safety committee consisting of Members from all site departments and Major contractors.

- The Site in-charge & Site Safety Manager shall be the head for the safety committee. The committee shall meet once in a month.

- Site Safety Engineer will be the Secretary of Safety Committee & he shall organize meetings.

The responsibility of the safety committee shall be the following
- Periodic site inspection with safety committee members.

- Following all the norms stated in the OHSE policy, Local Legal Requirements.

- Reporting all unsafe conditions / acts.

- To discuss all Action to be taken to prevent accidents.

- Minutes of safety committee shall be distributed to the members.

The safety committee agenda shall include (but not limited to).
a. First aid injury
1. Discussion of accidents record
b. Reportable injury
2. Safety inspection report
3. Safety training
4. Safety Activities.
5. Near Miss

6. SAFETY INSPECTION:

6.1 DAILY SAFETY INSPECTION:

Safety Engineer shall make an inspection of the site and find all unsafe conditions / acts and advise respective Supervisors / Engineers (Construction Dept) for taking corrective action in a time bound manner. All such actions shall be recorded.

- The Safety Engineer (Safety Dept) shall make an inspection report once in a week and daily safety report for both Safety Engineer and Safety Supervisor. The same shall be made with all entries by using a check list.

- Any problem regarding failure to rectify unsafe condition / action shall be highlighted to Site In charge for further action.

6.2. WEEKLY SAFETY REVIEW:

- The line Supervisors / Engineers (Construction) at their respective work place shall carryout regular safety inspection along with Site Safety Engineer.

- A register shall be maintained from the commencement of the project. This register shall include Electrical equipment, gas cutting torches, lifting machinery, welding machines, portable equipment and other accessories used in the site. The safety Engineer shall conduct inspection and verification of such records and equipment with concerned supervisor every week.

- Any contract equipment / Accessories coming to the site shall be checked by Supervisor & Site Safety Engineer.

- Walk through safety survey will be made by Site In charge, Customer Representative, Safety Engineer, Supervisors along with Contractor in the respective area for taking corrective action.

6.3. SAFETY INSPECTION BY SAFETY COMMITTEE MEMBERS: (ONCE A MONTH)

The site safety committee shall inspect all the activities in the project for identifying unsafe practices, safety hazards and safety violations.

- o Analyze safety performance at each work area and document any violation.

- o Observe the working conditions and the workers activities in all operations at the site.

o   Each violation will be reported to the supervisor who will be responsible for taking necessary corrective action.

o   Any dangerous situation shall be corrected immediately on the spot.

o   The findings of the inspection shall be made as a report and shall be discussed in the monthly safety committee meetings and for taking necessary action to prevent accidents.

7.   JOB SAFETY ANALYSIS:

- This procedure is used to review the methods followed in the job and to find out uncovered hazards in tools, equipment, processes and any change in the work procedure.

- All critical jobs shall be identified for conducting job safety analysis. JSA shall be prepared based on the input from crew members working on that activity with the help of safety department. The members consisting of Safety Engineer and Site Engineer. Safety & Construction Supervisors are responsible for conducting the job safety analysis and implementing safe work procedure derived from JSA. All work men and contractors are responsible for adopting the safe work procedure in doing their job.

- The Safety Manager will prepare and update the above for activities not covered.

- JSA for critical activity shall be reviewed by qualified senior person.

- JSAs with common activities will be covered by a standard set of JSAs that shall be kept easily accessible to the construction team. The Safety Engineer (Safety Dept) shall continually monitor that the procedure is being implemented.

- Sample JSA is attached in Appendix 1 for reference. This will be prepared by the SSM at site depending on the site condition.

8.   OPERATIONAL CONTROL:
8.1 SAFETY RULES AND REGULATIONS:

Following are the Rules and Regulations to be followed at the site:

✓   All workmen, employees and visitors shall wear all required PPEs at all times when they are at site (Helmet zone area) and shall wear their PPEs till the

Security check point. Free Helmet Zone Areas will be given by Site Safety Incharge, like Site office & shed etc, where hard safety helmet is not required.

✓ The security at the gate shall make the visitors entry in a register and issue safety helmet to the visitors.

✓ Drinking Alcohol is prohibited at site. Smoking shall be permitted in designated areas only.

✓ Do not spit indiscriminately at work spot.

✓ Do not tamper or remove the safety guards provided.

✓ Do not wear rings and jewellery when working with task specific works like electrical equipment, acidic areas etc.

✓ To maintain good housekeeping, dispose all litters and waste in the proper bins.

✓ Do not undertake any work, unless you are trained and authorized.

✓ Do not stand near unguarded edge, cut outs etc where you may fall from height. No unguarded holes and if someone notices any, they need to notify their supervisor so it can be immediately protected with a barrier.

✓ Always use safest route and do not make any short cuts.

✓ Persons working at height more than 1.8 meters level must wear safety harness and anchor it properly at all times, unless work is being performed from an approved platform that has been inspected and certified in writing by Safety department, with a copy on platform and in records, where no harness is required.

✓ Use appropriate PPE wherever required.

✓ Always adhere to local Laws / Rules / Regulations prescribed by Client, Local Govt. etc.

✓ Do not allow any child worker to work at the site. Child workers will be controlled at the entrance to the site. Small children who accompany their mother will not be allowed to enter the site.

✓ Workers shall use the sanitary facilities provided and those caught going randomly around the site will be disciplined and anyone caught doing so inside a building will immediately be removed from the job site.

✓ Workers should clean their area after completing the work and will never enter a work area that hasn't been cleaned up unless specifically directed by their immediate supervisor.

✓ Workers should never enter an area that doesn't have proper access unless directed by their immediate supervisor and they should only enter through establish proper access.

✓ Meals shall be taken in designated place only to prevent garbage laying around which causes rats.

✓ Removal of safety guards will be called out for hand tools as well as for all equipment

8.2 SAFE WORK PROCEDURES / WORK PERMITS:

Safe work procedure is to be followed for each work activity as listed below and proper SOPs shall be displayed at appropriate locations.

- Working at height (Refer Appendix 2)
- Scaffolding
- Ladder
- House keeping
- Electrical work (Refer Appendix 3 & 4)
- Welding and Gas cutting (Refer Appendix 5)
- Grinding
- Fire prevention & control
- Concrete & Masonry work
- Excavation work (Refer Appendix 6)
- Material transportation & handling
- Confined space (Refer Appendix 7)
- Plant & machinery
- Cranes (Refer Appendix 8)
- Lifting equipment – Forklifts and Man lifts

8.3. OPERATIONAL CONTROL PROCEDURES:

8.3.1. OCP FOR GRINDING OPERATION

- Grinding equipment like rinding machine, grinding wheel and electrical wires shall be thoroughly checked for the damage before starting the work. Work piece shall not be forced against a wheel. Wheel speed shall be checked before changing.

- Wear all the personnel protective equipment like Safety Shoe, Leather gloves, Leg Guard, Goggles, Grinding Face Shield and Safety Helmet while working.
- Always ensure the machine operating speed does not exceed the grinding wheel speed.
- Ensure all handles, guards and safety shields are in position before starting the grinding operation.
- Don't grind on the side of the wheels, unless the wheel is specifically designed for that purpose.
- Switch off all the electrical supply when not required.

8.3.2   OCP FOR WELDING OPERATION

- Welding cables, electrode holder, shall be thoroughly checked by Welder & Shop Incharge for any damage before starting the work.
- An unskilled labour will be available as a helper where welding is going on to watch for fire because the welder will not be aware of what is going on outside the hood.
- Wear all the personnel protective equipment like Safety Shoe, Leather gloves, Leather Apron, Leg Guard, Welding Head Shield, Full hand sleeve dress while working.
- Set the parameters for welding operation as required.
- Set the Electrode oven used for maintaining the temperature of electrode to around 150°C
- Electrode stubs shall be canned and disposed of properly and should not be thrown at site.
- Check the welding machine is properly earthed to prevent arcing.
- Check and maintain incoming electrical cable in good condition and make sure connections are tight
- Electrical equipment shall be inspected periodically and defects shall be reported to maintenance for rectification.
- Remove all inflammable materials from the location of welding to avoid fire and explosion
- Switch off all the electrical supply when not required
- Don't wrap cables around body

- Welding cable joints and connections shall be made with a connector specifically manufactured and insulated for connecting welding cables. No field splices or tape allowed on welding cables. Any welding cable with broken, cut or damaged insulation shall be removed from service immediately.

- Welding cables used for grounding shall have proper grounding clamp manufactured for that purpose.

- Hot Work Permits for Cutting and Welding will be issued by area only until the start of mechanical and electrical work after which the permits will be issued for isolated areas by everyone working in that area.

- Grounding lead clamp must be attached securely to the work piece being welded within a maximum distance of 2 meters from the welding taking place. Welding current from electrode to ground shall not pass through any motors, pumps, bearings, electrical equipment or other equipment that may be damaged by the current. Welding near or around any rotating equipment whether in service or not may require special procedures to prevent damage to equipment.

- Welding cables shall not lay in water or places that may become flooded with water.

- Fire blanket shall be used to prevent sparks/hot metal falling from height during gas cutting, welding and grinding. Fire blanket shall be used to cover equipment for protection from sparks/hot metal during gas cutting, grinding, and welding whenever necessary.

- A charged extinguisher shall be present at all times in the area where welding is being carried out.

8.3.3   OCP FOR GAS CUTTING OPERATION

- Gas cutting equipment like hoses, gas cutting torch and cylinders shall be thoroughly checked before starting the work.

- Wear all the personnel protective equipment like Safety Shoe, Leather gloves, Leg Guard, Gas Cutting Goggles, Helmet, Full Hand Sleeve dress while working.

- Ensure that the colour of LPG hose in Red and colour of Oxygen hoses in Blue.

- Cylinders, cylinder valves, hoses, regulators and gas cutting nozzle shall be kept away from oily or greasy substance.

- Always ensure that Heat / Spark / Flame are away from cylinders, articles, machinery, surfaces etc.

- No combustible substance like Paint drum, Thinner can and flaw check shall be kept nearby while gas cutting.

- Don't roll cylinders on floor; ensure use of trolley for movement of gas cylinders.

- Ensure the gas pressure for gas cutting to be around

- LPG and Oxygen cylinders shall always be kept vertical and tied off at all times.

- All cylinders shall have protective valve caps properly attached over the valves at all times when not in use.

- A charged fire extinguisher shall be available in the cutting area before any cutting is carried on.

- An unskilled labour shall be available with all cutters to keep an eye on the surrounding area for fires that might start while cutting.

- Cylinders shall not be stored in sea containers or any other such enclosed unventilated spaces.

- Hot Work Permits for Cutting and Welding will be issued by area, only until the start of mechanical and electrical work, after which the permits will be issued for isolated areas by everyone working in that area.

- Gas and Oxygen cylinders shall not be stored together when not in use, storage areas must maintain 6 meter min. distance of separation unless a fire wall, min 2 meter high with a fire rating of 30 min, is constructed between them.

- Cylinders shall not be transported with Gauges attached.

- Cylinders shall be protected from over head falling objects at all times.

8.3.4 OCP FOR MATERIAL HANDLING OPERATION

- Crane Chain slings & wire ropes shall be carefully handled and should never be overloaded and it shall not be attached to points that are not designed for lifting.
- All operations must be carried out as per coded signals given by the rigger to the crane operator. Hand signals shall be made available and the workers will be trained to use them. Crane operators will be handed with two way radio transmitters for communication.
- While operating at high angles, care shall be taken that the job does not hit the boom lattice.
- Riggers & operators shall carry a training certificate / badge so that they can be distinguished for signalling the operators.
- Always keep the hook at the centre while lifting the load. It shall not be suspended and travelled for more that 10 m and when transported, it shall be secured so that it does not swing wildly.
- The speed of mobile crane shall not go beyond 20 km / hr.
- The weight of the material can be judged before lifting and it shall be done by trained operators and riggers. The capacity of the cranes should be displayed in the machine.
- Don't make any knot in the chain sling.
- Proper equipment for handling & transporting material shall be used for that specific work for which it is designed for.
- All equipment & control switches shall be thoroughly checked before use.
- Use of wooden plank shall be ensured before placing the load on the floor / stand.
- Warning bells shall be ensured by the operator while travelling from one location to another location and on swinging any load near worker.
- Unauthorized persons shall not operate any material handling equipment.
- Do not keep the load hanging in chain slings for long time and left unattended.
- All material handling equipment shall be switched off when not required.

- Trained forklift operators shall be used for operating the forklifts. Clearance shall be analysed first and forklifts shall be used to move around the site.

- No suspended loads shall ever be swung over the heads of any workers and anyone involved in such incidence will be immediately suspended until further action can be determined.

8.3.5 OCP FOR CRANES & LIFTING EQIUPMENTS

- Before beginning any crane operation, the supervisor and operator should complete the pre-operation checklist

- When wind velocities are above 32 km/h (20 mph), the rated load and boom lengths shall be reduced according to manufacturer specifications. All lifts above ground level, must account for wind force, i.e., side loads, down drafts, etc. as applied to the load and boom.

- Shall ensure that the relevant certification is in order prior to use. In case of a lifting appliance such as crane, the certification should confirm that it has been tested and thoroughly examined by a competent person in the preceding 12 months.

- Before any crane in used, the operator shall examine the foundation, cables, drum, brakes, boom, guards, pins, sheaves, load hook, and wire line for defects. Any defects must be repaired before the crane is used.

- Signal man and operator jointly check load charts; confirm the boom length with the chart, establish load weight, and establish the boom angle.

- There is an area surrounding every power line that is referred to as the absolute limit of approach. It is strictly forbidden to move any crane boom or load line or load into this area unless the line has been de-energized or insulated.

9. NEAR MISSES & ACCIDENT REPORTING AND INVESTIGATING:

9.1 DEALING WITH NEAR MISSES & ACCIDENTS:

In case of Accident:-

- Near Misses shall be recorded and reported so that preventive measures shall be executed to prevent future accidents.

- First Aid shall be given to the accidentee and they shall be taken to the nearby hospital if needed at once.

- Persons with knowledge of first aid shall be made available at site by providing safety training to maximum staff as possible.

- Contact telephone number of hospital, fire service, police station, customer representatives, Site In charge & Site Safety Engineer shall be displayed at prominent places in the site, as per the Emergency Response Plan attached in Appendix.

- Workers shall be trained not to move the injured person, unless he is in imminent danger of further injury, before informing to the Safety team

9.2 REPORTING OF NEAR MISSES & ACCIDENTS:

All Near Misses & Accidents will be reported to Site Incharge, Safety Manager, Site Safety Engineer and the Site In charge shall take necessary action to give top priority to save the lives of accident victims as per Emergency Response Plan. If the incident involves any unsafe condition or could happen again, action shall be taken immediately to rectify the situation, action shall be taken to stop the activity and that particular labour will be terminated from site.

9.2.1 IN CASE OF REPORTABLE INJURY / ACCIDENT:

Site In charge shall follow all procedures as per The Factories Act 1948 & Local Factories Rules, as applicable to the customer, in case the injury is Reportable / Fatal. All accidents shall be reported to concerned at corporate office.

In case of Fatal / Reportable accidents

The Site In charge shall report the same to

1. Customer Management
2. Regional office (if applicable)
3. Corporate office

9.2.2 THE BASIC CAUSES OF ACCIDENTS IN CONSTRUCTION:

- Overhead operations in scaffolds

- Overhead Tower cranes and Material Handling equipment's

- Persons & tools falling from height

- Persons being struck or trapped by objects in motion.

- Persons handling objects in such a way so as to cause injury.

- People working in areas adjacent to operations where flying objects are present, without eye protection

- Person falling from edge of excavations or the edge of floors above ground level.

9.2.3   MAIN CAUSES OF CONSTRUCTION INJURIES:

- Fall from higher elevation

- Struck by falling material / equipment

- Caught in between material/ equipment

- Electrocution

- Cave-in (during excavation)

- Explosion/ Fire

- Eye injuries from concrete work, welding, cutting, grinding

- Asphyxiation due to Toxic gases

- Drowning

- Using improper hand tools & equipments

- Failure to wear their respective PPE's

- Failure to follow safety rules and regulations

- Improper practice when working with cranes, forklifts & man lifts

10   PERSONAL PROTECTIVE EQUIPMENT (PPE):

PPE is an important line of defence to avoid any injury due to accident. The Site Safety Engineer and Site In charge shall ensure the following

- All employees including visitors shall wear safety helmet in the site.

- High visibility vests shall be used for everyone working at site so operators can easily see where people are located.

- Persons working at height more than (1.8 meter level) must wear safety harness, anchored properly and helmet.

- Site In charge shall enforce the usage of PPE at the site. The safety Engineer shall give proper training to all employees for proper use and care of PPE.
- Safety Devices like Fall Arrestors, Safety Nets, Double lanyard safety belts, Flashback arrestors in gas cylinder sets etc. must be ensured.

10.1 LIFE SAVING DEVICES:

- Safety helmet
- Safety Harness
- Fall arresters
- Safety Net
- Life rings around water holes (except water reservoir)
- High visibility vests
- Handrails
- Barricades

SAFETY HARNESS

- For the work more than 1.8M height, 100% tie off is mandatory at all times
- Life line anchoring point fixed at higher elevation
- Maximum free fall 1.2M
- Confirming to IS 3521-1989

LIFE LINE

- It is otherwise called as static line
- For anchoring safety harness while moving at height
- Life line length- 3M (Length of line from safety harness and the anchorage)
- Minimum Factor of safety of life line = 10 required and shall be tested with a dead weight with the safety engineer

- To facilitate free movement of tie rope & safety harness.

- Check after every shift for damage and it shall be recorded.

- All parts of the Life Line shall be protected from subsequent damage including the possibility of being damaged by sharp edges, chemicals, fire, etc.

SAFETY NET

- Protect person from falling materials

- Mesh size 19.6*19.6mm

- It should have adequate Breaking strength as displayed in the net

- Size of safety nets standardized is 10*5M distance of fall should not exceed 6mtrs.

- Storage: Protect from damp and heat. If wet, should be dried naturally. Stack on timber

11. FIRST AID / MEDICAL FACILITIES:

The site shall have a first aid room equipped with all required Medicines and a Doctor working 6 days a week 8 hrs per day. The Site Safety Manager & Site In charge shall make arrangements for replenishing all required medicines and maintain the First Aid Box in good condition always. Joint monthly inspections of the medical facility shall be performed by the Safety Manager and at least two Customer senior staff representatives. At that time the performance of the facility shall be reviewed and action will be taken to continually improve the service. All workers who are ill when they report to work plus all workers that get ill during a shift must report to the medical facility for examination. Routine health exams shall be performed on site on each employee at least every 3 months so tests can be run for the potential to carry communicable diseases. The medical staff, under the supervision of the Doctor, will provide monthly reports describing the treatments provided as well as recommendations for providing additional equipment or medicines that would improve the service provided.

12 SAFETY RECORDS:

The following safety records shall be maintained at the project site and continually reviewed and assessed throughout the project duration. A copy holders list will be in the index page of the folder.

1. Accident & Near Miss Reports

2. Accident & Near Miss Register

3. Accident & Near Miss Investigation Report

4. Safety committee Minutes

5. Safety Inspection by Safety Engineer

6. Safety training Records

7. Unsafe condition / Action Report

8. Permit to work

9. Job Safety Analysis Report

10. Monthly safety Report

11. Accident Statistics

12. Medical Report

13. Test certificates (if applicable)

13 SAFETY PERFORMANCE MEASUREMENT:

Records of frequency rate and severity rate shall be maintained at site.

$$\textbf{Frequency Rate} = \frac{\text{Total number of Reportable Accidents}}{\text{Total man hours worked at the site}} \times 10^6$$

$$\text{Severity Rate} = \frac{\text{Total mandays lost due to Reportable Accident}}{\text{Total man hours worked at the site}} \times 10^6$$

(a) Reportable Accidents:-

It is the accident causing personal injury, making disability or absence from duty for more than 48 hours, damage to equipment's, vehicle and parts.

(b) Total man hours worked at the site:-

This is the total man hours of staff and workmen including all contract workmen (Over time also to be include).

(c) Man days lost:-

It is the disability period in days of all the injured persons arising out of Reportable accidents. Man days lost due to fatal accidents and permanent disablement will be calculated as per safety code. (Indian Standards)

This safety performance shall be discussed during site safety committee meeting and suitable measures will be taken to improve the safety performance.

14. EMERGENCY PREPAREDNESS PLAN:

- The Emergency response plan is given in the appendix and shall be updated to be current.

- The Organization shall also periodically test such procedures.

- The Site in charge is the responsible person, and he shall act as the evacuation coordinator who will perform the following functions.

  ➢ Determine whether there is a need for evacuation.

  ➢ Approve the evacuation procedure and locations of Safe / Emergency assembly points.

  ➢ Direct to activate evacuation alarm, and follow evacuation activities. Alarms will be located around the site in sufficient numbers that all employees in the project site can hear the alarm when activated. Alarms will be tested weekly.

  ➢ Instruct supervisors and Engineers to notify the contractors and project officers about the evacuation decision.

  ➢ Maintain the current list of emergency services like Ambulance, Doctor, Hospital, Fire Department, Police Dept, etc. to be called for assistance.

  ➢ To inform Client about the nature of emergency and all events.

In case of any major emergency the following procedure to be followed.

INITIAL RESPONSE:

First person to recognize an emergency is expected to immediately contact the following people:

1) Caller's first contact is Mr. _____ who can be contacted by Phone at _____.

2) Caller's second contact is Mr. _____who can be contacted by Phone at_____.

3) Caller's third contact is Mr. _____ who can be contacted by Phone at_____.

INITIAL RESPONDER ACTIONS:

The initial responder will then contact the required emergency services as listed below: List of emergency contacts:

- Male Nurse -
- Ambulance - 102
- Hospital -
- Police - 100
- Fire Department – 101

After the proper emergency services have been called, the initial responder is then responsible to:

See the scene of the emergency is secured, either directly or by assigning the task to another responsible person. The first responder then, is responsible to direct the required responsible people to meet the emergency services previously contacted and direct them straight to the scene of the emergency.

Assess the scene of the emergency looking for injures and confirming that a responsible person is taking care of all injured personnel.

After addressing all injuries, the initial responder shall assess the scene for property damage and the actions required to minimize any further property damages.

After actions have been taken to protect life and property, the initial responder shall contact the following responsible parties:

1. EPC Site In Charge-Mr. _____
2. Customer Site In charge Mr. _____
3. Customer Factory manager Mr._____

RESPONSE / PREPARDNESS TEAM:

Response Teams will be formed and trained with one response team on duty at all times that work is ongoing at site. The Response Team on duty at any given time, shall be clearly posted at all entry points to the project such that every worker entering the site to start his shift will be aware of all team members. All worker communications will be direct to the worker's immediate supervisor.

The initial site Response Team will be headed by the Site Safety Manager.

As the Response Team Leader, the Site Safety Manager will be the first responder and will be supported by the Assistant Team Leader, Mr. _____. The other Response Team members will include M/s _____.

Mr _____, as the Assistant Response Team Leader, will assume the duties of the Response Team Leader should the Site Safety Manager not be available. The Response Team shall meet at least once a week, to review all recent changes to the plant and procedures and to participate in a weekly training in the different aspects of properly responding. This response document is also to be updated weekly at this meeting and a current copy shall be posted all around the site in the local language as well as English so everyone is aware of the current proper response to an emergency.

CONTROL OF SCENE:

During an emergency, Mr. _____ has the overall responsibility for controlling the site, both at the scene of the emergency and the other plant areas, per the Site Specific Safety Plan. The Line Supervisors in each area have the responsibility to control the workers in their charge to ensure that ongoing works continue and are not affected by the emergency situations that can develop.

AMBULANCE:

Provision of ambulance at site will be as per the contract. The Customer ambulance will be the primary vehicle used to transport all injured personnel to the designated Hospital for treatment. Prior to removing the ambulance from service at site, Customer shall inform EPC contractor so a substitute vehicle is available at all times. If the ambulance available at any given time does not have four wheel drive, EPC contractor shall assure that at least one four wheel drive vehicle is available at all times to transport the injured from the remote site to where they can be loaded into the ambulance. The initial responder shall have the responsibility to see that the designated local Hospital is advised that the ambulance will be leaving our site for the hospital with an injured person. The response team shall then advise the hospital of the condition of the injured, immediately after the person leaves the scene. An emergency ambulance or an additional ambulance is available 24 hours a day by calling 102 on any mobile phone.

FIRE RESPONES:

Customer has the responsibility to provide the project fire tender and fire station. Until such time that customer has furnished this equipment and facilities, EPC contractor will furnish a water tanker with a committed and operational pump dedicated to responding to a fire. The water tanker shall have a tractor connect to it at all times and may be used for other construction activities as long as it is clear to everyone that in case of a fire this tanker immediately responds. During off hours the water tanker shall be full and left connected to the tractor for any required responses during non working hours.

14.1   SAFETY MANAGER / ENGINEER:

He shall do the following when directed,

- To alert first aid, fire brigade to be ready for action.

- Arrange persons trained in first aid to attend immediately the injured & transport them to first aid station / Hospital.

- Obtain outside help if necessary – for fire service, ambulance service & medical help.

- Shall act as a liaison between the evacuation coordination team and first-Aid crew.

14.2   SITE ENGINEERS / SUPERVISORS:

They shall arrange for

- Evacuating all persons and assemble them in Safe / Emergency assembly point.

- Collect the head count of the personnel reported in safe assembly point and forward the same to the concerned site in charge / incident controller for conformation.

- Check the missing persons if any and report to the evacuation coordinator.

14.3   EMPLOYEES:

They have to follow

- Upon hearing the alarm signal, stop all their work and proceed quickly to safe assembly point / area.

- After reporting to their immediate supervisor, they have to move directly to the assembly area and wait for instructions from their supervisor as to when to leave that area and where to go.

14.4   ASSEMBLY AREA:

- Assembly area is defined as a location to assemble during emergency for proper evacuation. This location will be decided by the Safety Manager and then placed on a map that will be distributed and posted around the site.

- The immediate supervisor of each crew has to account for all his workers when the emergency siren go on. Once he has accounted for all his crew then he himself will evacuate to the assembly area and report on the presence of his crew to the person incharge of the Assembly Area at that time.

- If necessary, the Person in Charge of that Assembly Area will give further instructions as to the need for assistance in searching for any missing employees but no one shall leave that Assembly Area without the approval from Person in Charge.

- The assembly area shall be located near to the gate and close enough to work areas for easy access.

14.5   TRAINING:

- Training of all employees including contract workers shall be given on Emergency Plans by Site Safety Manager.

- Necessary review in the procedure shall be made once a month.

- The Emergency Plan with work response plan shall be included in training for new employees in the induction safety training.

- Notice shall be posted for the information of all employees about emergency plan and work response at site.

- The training and posting will be available in local language, Hindi and English.

- Planned Mock drill should be conducted

Display of Emergency Phone Numbers:-
The following Personnel contact numbers shall be displayed at a few locations.
EPC Contractor SITE
Site Incharge:

Safety Manager / Engineer:
Regional Office:
Corporate Office:

CUSTOMER PHONE NOs

Project Manager:
Safety Officer:
Security:
Fire Service:
First Aid centre:

Outside Agencies:

Hospital:
Ambulance:
Police:
Fire Service:
DCIF / IOF:

15 SAFETY IN EXCAVATION:

- Soil to be tested to find out presence of any harmful gases

- Necessary arrangements are to be made to remove/ dilute such gases.

- To check for presence of underground installations, like cables, pipelines etc.

- Warning signs / Caution boards are to be displayed with proper illumination, to have clear visibility during night.

- Blastings for excavation are to be carried out as per explosives handling and storing rules.

- Controlled blasting is to be carried out with cordoning off the area to prevent unauthorised entry.

- No loose Materials shall be kept very close/around excavated area.

- Proper timbering and shoring is to be ensured to prevent the loose soil sliding while the workers are working inside the pits.

- Barricading is to be provided for the excavated pits more than 1.5 metres depth.

- Movement of vehicles very close to the edge of the excavated area is to be prohibited.

- All safety equipment and signage shall be used at site like Safety Helmet, Warning Signs / Caution Board, Barricading Tape, Safety Shoes

16   SAFETY IN SCAFFOLD:

- Scaffold is a most commonly used access and platforms for working at height.

- Scaffold is a temporary structure consisting of stairs, ladders, walkways, guard rails and toe boards.

16.1   CAUSES FOR SCAFFOLD HAZARDS

- Failure of Scaffold foundation due to
  - ➢ Soil washed away/ uncompacted soil
  - ➢ Excessive load
  - ➢ Improper bracing and tying to the structure
- Inadequate design considerations
- Inferior quality of materials used
- Improper alignment
- Use of combination of improper materials.
- Insufficient and unguarded working platforms.
- Vehicles plying at site hitting from below.

16.2   HAZARDS ASSOCIATED WITH SCAFFOLDS ARE

- Fall of persons and materials from working platform.
- People below struck by falling materials from height.
- Collapse of scaffold causing damage to property and injuries to persons.

16.3   SAFETY PRECAUTIONS

- Factor of Safety at least 4 to be considered in design.

- Proper and sound materials are to be used.

- To be erected by experienced and trained persons under the supervision of a competent person.

- Foundation to be on firm base with adequate bracings at all levels.

- Metal scaffolds to be erected away from Over Head Transmission lines as per norms.

- Gangways / Ramps to be minimum 500 mm wide with 1:1.5 slope maximum.

- Working platforms with adequate width, guardrails and toe boards to be provided.

- Each plank must overlap the support by a minimum of 6 inches but should not extend beyond the support more than 18 inches.

- Planks must extend over the supports and overlap each other by at least 12 inches.

- If minimum clearance between the power lines and scaffolds is not possible, the power lines are to be de-energised.

- Use of electrically conductive tools or materials while working on the platforms / scaffolds must be restricted near the overhead power lines.

- Planks / platforms should be properly secured to scaffolds to prevent uplift or displacement because of high winds or other job conditions.

- Not more than one person should be allowed to stand on one plank at one time.

- Scaffolds should never be used as material hoist towers or for mounting derricks unless the assembly is designed for the purpose.

- When necessary to prevent danger from falling objects, scaffolds should be provided with adequate screens.

- Every scaffold should, before use, be examined by a competent person to ensure more particularly:

  a) That it is in a stable condition;

b) That the materials used in its construction are sound;

c) That it is adequate for the purpose for which it is to be used; and

d) That the required safeguards are in position.

16.4 SCAFFOLDS SHOULD BE INSPECTED BY A COMPETENT PERSON AND TAGGED CLEARLY:

a) Once a week.

b) After every spell of bad weather and every prolonged interruption in the work.

c) Scaffold parts should be inspected on each occasion before erection.

d) No scaffold should be partly dismantled and left so that it is capable of being used, unless it continues to be safe for use.

Handrail

- Height of hand rail min. 42".
- Height of the toe board min. 4".
- Working Platforms with Hand Rails shall be provided

17 LADDER SAFETY:

- Ladder should be Sound in construction
- It should not be slippery
- No Make shift ladder
- Proper ascending and descending method
- No Carrying of object in hand
- Firm ground and strong top support
- Portable ladder max. 6M in length
- Permanent Stair case – preferable
- One Person at a time
- Ladder should be lashed at the top

- Extended 1M above the platform

- Rungs should be free from defects

- Interval between rungs max. 30cm.

- 75° angle with horizontal

- Anti-Skid hold at bottom

- Ladder shall be lashed to a fixed Structure

18   ELECTRICAL SAFETY

- ELCBs shall be provided for all the electrical systems and checked for their healthiness periodically by Safety, EE and Concerned contracting agency safety officer.

- Ensure all electrical cables are free of joints by regular inspections and continuous monitoring.

- Insulation healthiness of cables is ensured by continuous checking.

- Lock Out Tag Out procedure shall be followed in case of repair. If anything is defective or any electrical line is under repair, a tag shall be displayed and locked in the main switch displaying the area of problem and the person who is working on the same. So any person can see that the particular work is under repair and it is locked due that defect. The tag should contain the details of the person signing the tag, his staff number, his mobile number & name of the person. Use a standard danger tag for labelling. Destroy the tag when it is removed and a new tag will be used for other work.

- Regular joint electrical safety inspections are conducted at all the electrical systems.

- Use of correct rated fuses is always ensured.

- Check for bare conductors & avoid joints.

- Check for broken plug & socket and replace at once.

- Check all switch boards for proper connections.

- Wear all personnel protective equipment like Safety Helmet, Safety Shoe, and Electrical Safety Gloves while working.

19 SAFETY PROCEDURES FOR CRANES AND LIFTING EQUIPMENT

- The load on the crane shall never exceed the safe allowable working load specified by the manufactures of the crane.

- Damaged and worn lifting tools and tackles such as hooks, rings, eye bolts, chains and wire or fibre ropes and slings shall be immediately removed from service.

- Cranes and lifting equipment will only be operated and used by trained and experienced personnel. No crane or lifting device will be used on site unless it is in good condition and safe for use.

- No lifting operation will be carried out until a proper and thorough assessment has been made, having taken into account, among other things, the positioning of crane outriggers, slinging arrangements and banks men control.

- Cranes and lifting equipment will be clearly marked as to their lifting capacity and not used beyond that capacity. Training will be given to crane drivers and riggers, stressing the necessity for a cautious approach to the lifting operations.

- The crane operators will not operate the crane until all persons concerned have been instructed as to the work to be done.

- A qualified signal man shall work with the crane operator. Standard signal must be used. Normally, the signal man must give all signals, but the operator should obey the emergency stop signal given by anyone.

- All cables should be changed when appreciable wear or corrosion is shown. Cables should be inspected daily. Sling found to be defected shall be discarded. The date on which the sling is placed in service should be stencilled on the metal connector.

- The operator must properly secure the crane and boom before or latches shall be used on crane hooks each time is lifted.

- Load limits for all crane and "stiff legs" shall be posted in clear view of the operator. Boom angle indicators will be permanently attached to the boom to show operating radius.

- Maintenance inspection will be made monthly on cranes "stiff legs", and their wire lines.

- The lift area would be cordoned off.

- No lifting would occur when conditions exceed acceptable limits (which depend on the type of crane / lifting equipment used for that particular work).

20 SAFETY PROCEDURES FOR WORKING AT HEIGHT

- Nobody shall be allowed to work at height (> 1.8 meter) without valid work permit and site supervisor to keep the permit while working.

- Safety Harness / 100% tie off shall be used by all workmen / supervisors, working at height.

- Sufficient access ladders, work place scaffolding, work platforms shall be arranged properly.

- Only trained man power shall be used.

- Persons working at >8 meter height are to be medically fit prior to start of work.

- The Tools / Materials taken to work place at height shall be kept properly in a holder and shall be brought down after work is over with proper checking.

- The concerned supervisor must be available at site during the period of work.

- Electrical power supply, if it concerns the place of work, shall be isolated.

- All personnel protective equipments like Safety Harness, Life Line, Fall Arrester, Safety Net, Double Shock absorbers, Full Body Harness, Helmet shall be used while working.

21 SAFETY PROCEDURES FOR HOUSEKEEPING

- The general site shall be maintained in a clean and tidy manner.

- Access ways and exits from the work place shall be kept clear of equipment, materials and garbage.

- As materials are unpacked, packaging materials, especially those of combustible nature shall be removed from the work site.

- Before entering contract, for the removal of solid wastes from site, with service contractor, verify the accreditation for such purpose and make sure that wastes are responsibly disposed of.

- The housekeeping check-list (Page 29) will be checked periodically.

22 SAFETY PROCEDURES FOR FIRE PREVENTION AND CONTROL

- Smoking is permitted only in designated areas. Matches/Cigarettes must be extinguished and disposed off in approved containers.

- The amount of flammable liquids/gases kept at the work area should be minimised.

- Containers of flammable liquids should be closed when not in use.

- Spills and indications of excessive flammable vapour/gas concentrations should be reported immediately.

- The necessary permits when performing hot work should be obtained.

- Materials and equipment should not be allowed to block the access to extinguishers and fire protection hoses, hydrants, and standpipes.

23. SAFETY PROCEDURES FOR WORKING IN CONFINED SPACES

- Confined space is an area which has a limited means of access and exit and restricted natural ventilation.

- (A sign in and sign out procedure for entry along with manned hole watch, for confined space where a permit is required.)

- All work performed within a confined space must be covered by a permit to work.

- Breathing apparatus will be worn unless the area is adequately ventilated and no substances are present that will generate dangerous fumes.

- No spraying, painting or coating of substances hazardous to health is to be undertaken in any confined space unless adequate precautions are in place to eliminate the health risk.

- No smoking, naked lights, torches, arcs, flames or other source of ignition is to be allowed within a confined space unless the atmosphere has been tested and proven safe.

- Adequate means of access and exit will be provided for all confined or enclosed spaces.

24. SAFETY PROCEDURES FOR CONCRETING AND MASONARY WORK

- Stability of shuttering work shall be checked before starting concreting work.

- PPE for the all the workers should wear the location for the work. (Like gumboots, rubber hand gloves)

- Concreting area shall be barricaded, if pouring at height / depth.

- Vibrator hoses, pumping concrete accessories shall be kept in healthy condition.

- Pipelines in concrete pumping system shall not be attached to temporary structures such as scaffolds and form work support as the forces and movement may affect their integrity.

- Safety cages / guards around moving motors / parts of concrete mixers shall be in place.

- Concrete mixers shall be provided with hoppers.

- Concrete mixers shall be inspected for their condition at start of work.

- Concrete mixers shall be maintained well so as not to generate excessive noise.

- Earthing of electrical mixers, vibrators etc. shall be done and verified.

- Personal protective equipment such as gloves, safety shoe and safety helmet shall be used while dealing with concrete, and nose mask shall be used while dealing with cement.

- Cleaning of rotating drums of concrete mixers shall be done from outside. Lockout devices shall be provided where workers need to enter.

- Adequate lighting arrangement shall be ensured for carrying out concrete work during night.

- During pouring, shuttering and its supports shall be continuously watched for defects.

APPENDIX – A. SAFETY INDUCTION TRAINING

Date: Time:

EPC Company Employees   Contractor Employees

SL.NO	Name	Designation	Signature
1			
2			
3			
4			
5			
6			
7			
8			
9			
10			

Safety Engineer

APPENDIX – B. ACCIDENT REPORT

Date:
Name of the injured person:
Age:
Address of the injured person:
_____

Contractor Details:
(Name & Address)
_____

Location of Accident:
_____

Date and time of Accident:
_____

Time of starting of work:
_____

Investigation Details:
_____Part of
the body injured:
_____

Witness:

Sl. No	Name	Designation
1		
2		

_____

Copy to:
Site In charge
Site Safety Engineer
Signature of Supervisor / Engineer

SAFETY FIRST ALWAYS

Date:

APPENDIX – C. ACCIDENT INVESTIGATION FORM

Accident Information				
Day & Date of Accident	Time	Location	Shift	Contractor
			1 2 O O	

Injured Person:

Name	
Age	
Job title	
Length of Employment at Company	
Length of Employment at Job	
Supervisor Name	
Employee Classification	Full Time Part Time Contract Temporary
Witness to the accident	
Injured part of body	

Nature of Injury Treatment				Treatment
Strain/Sprain	Laceration / Cut	Internal	Others (specify)	First Aid
Fracture	Scratch/ Abrasion	Foreign Body	Bruising	Outpatient at Hospital
Chemical Reaction	Amputation	Dislocation	Burn/ Scald	Hospitalization

Property Loss:

Property, Equipment, or Material damaged	Describe Damage:

Accident Investigation:

Where it happened

When it happened

Whom it happened

What was the reason?

Why it happened
How it happened
(Investigation)

Root Cause Analysis:

Unsafe Acts	Unsafe Conditions	Type of Injury
Improper work technique	Inadequate Training	Caught between object
Safety rule violation	Congested work area	Hit by object
Improper PPE or PPE not used	Hazardous substances	Hit against object
Operating without authority	Fire or explosion hazard	Fall of object
Failure to warn or secure	Inadequate ventilation	Fall of person
Operating at improper speeds	Improper material storage	Flying object
By-passing safety devices	Defective / Improper tool or equipment	Fire Injury
Horse Play	Insufficient knowledge of job	Electric shock
Improper loading or placement	Slippery conditions	
Improper lifting	Poor house keeping	
Servicing machinery in motion	Trip Hazard	
Unnecessary haste	No / Inadequate guarding of hazards	
Unsafe act of others	PPEs not available	
Other*:	Other#:	Other^:

Root cause of the accident

Recommendation for Prevention of Accident:

Sl. No.	Recommendations	Responsible Agency	PDC
1			
2			
3			

Conclusion:

Investigated by

Site Safety Engineer

Copy:

Concerned Supervisor / Engineer --- for Necessary action through' Site In charge

APPENDIX – D. UNSAFE CONDITION / ACTION NOTICE

DT:__/__/____

TO

Please have the following unsafe conditions corrected which was observed by the undersigned on _____ at _____hours in the section_____

at_____

_____

_____

_____

_____

_____

Please sign and return the form to safety cell within 10 days duly indicating the action taken on the recommendations made

Safety Engineer

APPENDIX – E. SAFETY INSPECTION

Name & Address of the Site: Date of Inspection:

Unsafe Condition & Action Noticed:

Signature of Safety Engineer

Report Received By:

Remedial Action taken:

Any other Remarks:

Signature of Site Engineer

APPENDIX – F. ROUTINE SAFETY INSPECTION CHECKLIST

Date:

Sl. No	Control	Subject Description	OK	Improvement	
				Needed, Action By	Not - needed
1	House keeping	1) Clear access to m/c s and switch Boards. 2) Clear walk way /gang ways are maintained 4 feet /1.2 meter width 3) Provided storage racks for tools, Raw materials parts etc. 4) Clean up oil spills or other slippery Condition. 5) Chips bins / Dust bins near to machine 6) Empty (oil /grease /paint) drums are not used as Work platform			
2	Passage way & Barriers	1) Passage way for clear movement of materials & People. 2) Proper hand rails and barriers near to dangerous Machines and process. 3) Electrical cable not crossing the Gang ways and roads.			
3	Cranes	1) Name plate details of cranes and Certification for Safe working load. 2) Lifting Tackles are certified for its Rated Capacity. 3) Proper functioning of the protective Devices. · Hoist limit switch. · calling bell. · Free rotation of hook · Hoist wire rope condition · Over load relay			

Sl. No	Control	Subject Description	OK	Improvement	
				Needed, Action By	Not - needed
4	Material unloading	1) Materials scattered /unwanted material placed. 2) Proper packing to the loads for easy and safe handling. 3) Material obstructing path way/vehicle movement. 4) Material stacking in order and tightly packed. 5) Free space is available for placing materials and components at the site.			
5	Safety Posters	Available at respective areas at the site.			
6	Safety slogan	Available at respective areas at the site.			
7	Uniforms	Proper uniform and safety shoes for all 1) Employees & 2) Contract Workmen			
8	Protective Equipment	Availability of Personal protective Equipment · Safety goggles · Ear plug · Helmet · Face shield · Hand gloves · Safety Harness			

Sl. No	Control	Subject Description	OK	Improvement Needed, Action By	Not - needed
9	Facilities For 1)Employees & 2)Contract Workmen	1) Toilets, sufficient in number and near to work area. 2) Hand wash facilities with soap & hand cleans 3) Regular cleaning of toilets and urinals 4) Clothes are not hanging in shop floor 5) Locker rooms for changing clothes and Rest rooms for workers. 6) Tool box, cupboards are placed in identified location in shops			
10	Health	1) Availability of first aid equipment and qualified first aid persons. 2) Health examination to of Employees & Contractor workers. 3) Ready access to treatment by a physician /nurse. 4) NDT activity not carried out in open /road, which cause harm to employees.			
11	Environment	1) Noise 2) Lighting 3) Ventilation 4) Greenery condition 5) (Plantation of trees for improvement) 6) Any specific pollution control measures			

REMARKS: -

APPENDIX – G. BUILDER HOIST INSPECTION REPORT

Name of the Site:
Job No. :
Inspected by:
Date:

Sl. No.	Points	Observation	Measures
1	Top Limit Switch		
2	Condition of Structural Members		
3	Size and condition of Ropes		
4	No. of Load lines		
5	Stability of Mast		
6	Brake		
7	Guards of moving & rotating parts		
8	Load carrying capacity		
9	Load test details		
10	Operators Fitness		
11	Fire Extinguisher in operator's cabin		

Signature &   Safety Engineer   Site Engineer
Date

APPENDIX–H.

VEHICLE & EARTH MOVING EQUIPMENT INSPECTION CHECKLIST

Name of the Site:

Job No. :

Inspected by:

Date:

Sl. No.	Points	Observation	Measures
1	Engine condition		
2	Clutch / Brake		
3	Hydraulic System		
4	Guards / Covers/ Doors		
5	Fastener lock pins / Keys		
6	Horn / Reverse horn / Lights		
7	Indicators / Wiper blades		
8	Operator Fitness		
9	Tyre Pressure		
10	Condition of battery & lamps		
11	Operating levers / Steering		
12	Gauges and warning devices		
13	Fire extinguisher provided		
14	Side mirror		
15	Seat belt		
16	Any other points		

Safety Engineer   Site Engineer

APPENDIX–I.

ELECTRICAL SAFETY INSPECTION REPORT

Name of Site:

Job No:

Region:

Area:

Date:

To:

CC:

Sl. No	Points	Present Condition	Safety Measures to be taken	Remarks
1	Earthing			
2	Neutral Earthing			
3	ELCB			
4	Insulation			
5	Cable Layout			
6	Protection from water			
7	Lightening Arrestor			
8	Plug Tops & Cable Joints			
9	Installation Cover & Seals			
10	Electrical Safety at Installation containing inflammable & explosive substances			
11	Safety Guards			
12	Others (Specify)			

Site Safety Engineer   Site Electrical engineer

APPENDIX – J. JOB SAFETY ANALYSIS

JOB SAFETY	ANALYSIS JSA No.	Date

SPECIFICATION:

Activity Break down the task in to steps	Hazard List the hazards that could cause injury	Remedial Action Control measures to minimize risk of injury	Remarks

Safety Engineer   Site Engineer

CC: Site In charge

APPENDIX – K. MONTHLY SITE SAFETY STATISTICS

Name of Site:

Job No:

Region:

Month:

Date of commencement:

Date of closing:

Total Man hours Worked during the Month

Sl. No.	Description	No.	Man hours worked	OT performed	Total
1	Company Staff				
2	Departmental workmen (Including regular supply)				
3	Sub contractors workmen (Including security personnel)				
	Grand Total of Man hours Worked During the Month				

Signature

ACTIVITY CHECK LIST - STRUCTURAL FABRICATION & ERECTION

Date:___/___/____

Sl. No.	Activities in Structural Fabrication & Erection	Observation	Remedial Measures
	All Electrically operated equipment has		
1	Proper Earthing and connected through ELCB.		
2	Safety guards for drilling & grinding machine are in position.		
3	Use of Scotch block / wedge on wheels of trailers during unloading of material.		
4	End stoppers fixed and maintained for rail mounted gantry cranes and limit switches are in operating condition.		
5	Checking lifting tool &tackles before use.		
6	Precaution during slinging on sharp edges.		
7	Signalling to crane operators by one person at a time.		
8	Withdrawal of persons beneath suspended load.		
9	Cordoning on all sides displaying Red Flags/Taps/Light and warning signs.		
10	Load to be lifted is properly ascertained to identify centre of gravity etc.		
11	Clear passages of men, posts, material etc and easy access for to move with suspended loads.		
12	Proper tag line is used for guiding lifting loads.		
13	Guy ropes are used and secured during and after erection of heavy lift.		
14	Proper sequence of erection is followed.		
15	Wire ropes are maintained and its safe working load inscribed.		
16	Adequate illumination provided.		

Safety Engineer   Site In charge

| | **ACTIVITY CHECK LIST - GAS CUTTING & WELDING** | | |

Date:___/___/____

Sl. No	Activities in Gas Cutting & Welding	Observation	Measures
1.	Storing of gas cylinder like DA, Oxygen full & empty etc.		
2.	Proper handling of gas cylinder.		
3.	Condition of Regulator, hose, torch etc.		
4.	Welding generators/transformers condition and its proper Earthing.		
5.	Condition of welding cable and joints.		
6.	Electrode holder.		
7.	Area free from combustible materials.		
8.	Cordoning when welding/Gas cutting is in progress at height.		
9.	Provision of fire extinguishers.		
10.	Smouldering fires are rigorously extinguished after day's job.		
11.	Stacking of cylinders not near live wires, battery charging rooms/oil rooms.		

Safety Engineer: Site In charge

ACTIVITY CHECK LIST - ELECTRICAL WORKS

Date:___/___/____

Sl. No	Activities in Electrical Works	Observation	Measures
1.	Earthing of electrically operated Equipment.		
2.	Provision of Shed/ canopy/ cover of distribution board and sub-distribution board.		
3.	Insulation of cables and joints.		
4.	Cable laying above 7' from ground level.		
5.	Fire Extinguishers in main distribution board room.		
6.	Periodical checking of portable tools.		
7.	Use of ELCBs.		
8.	Men working, Don't switch ON board and other related warning boards and Tags.		
9.	Insertion of loose wire and sockets.		
10.	Use of proper plug and sockets.		
11.	Job safety Analysis for shut down jobs and its proper action.		
12.	Permit to work.		

Safety Engineer: Site in- charge:

ACTIVITY CHECK LIST - HOUSE KEEPING

Date:___/___/____

Sl. No	Activities in House Keeping	Observation	Measures
1.	Material stacking and storing.		
2.	Working / moving area clean.		
3.	Access / Main Approach / Passages free from Obstacles.		
4.	Cordoning / covering of pit, vat, machine foundation etc.		
5.	Displaying of Red Flags / Tape / light.		
6.	Removal of unwanted materials like excavated Earth debris etc.		

Safety Engineer: Site in- charge:

ACTIVITY CHECK LIST - FIRE PREVENTION/PROTECTION

Date:___/___/____

Sl. No	Activities in Fire Prevention/Protection	Observation	Measures
1.	Combustible material away from source of heat / fire.		
2.	Provision of fire extinguishers and its maintenance.		
3.	No smoking Board / Caution Board displayed.		
4.	Stacking / Storing of different type of combustible materials.		

Safety Engineer   Site In charge

ACTIVITY CHECK LIST - WORKING PLATFORM

Date:___/___/____

Sl. No	Activities in Working Platform	Observation	Measures
1.	Should not be less than 600 mm Wide.		
2.	Guard rails adequate check. a) Top rail at least 910 mm height. b) No gaps greater than 470 mm		
3.	Border free of defects check. a) No gaps. b) Adequate supports c) No risk of trips. d) Properly secured / tied.		

Safety Engineer   Site In charge

ACTIVITY CHECK LIST –

Inspection & Activities Personal Protective Equipments

Date:___/___/____

Sl. No	Inspection & Activities in Personal Protective Equipments	Observation	Measures
1.	Helmet and Footwear usage.		
2.	Safety Harnesses as required and secured properly.		
3.	Use of safety harness & Helmet while working at height.		
4.	Safety goggles during welding / gas cutting / Grinding etc.		
5.	Condition / Maintenance of safety appliances.		
6.	Use of body guards, gloves etc.		

Safety Engineer: Site in- charge:

ACTIVITY CHECK LIST - MISCELLANEOUS

Date:___/___/_____

Sl. No	Miscellaneous	Observation	Measures
1.	First Aid box with proper medicines and its maintenance.		
2.	Validity data of medicines.		
3.	Illumination.		
4.	Safety board and safety promotional materials. a) Posters b) Stickers		
5.	Accident Report Form.		
6.	Safety Awards / Promotional activities.		
7.	Reporting system with HO / Regional office.		
8	Arrangement of drinking water and sanitation.		
9.	Provision of emergency Vehicle.		

Safety Engineer: Site in- charge:

APPENDIX - 1 - JSA SAMPLE

S. No	Activity	Hazard	Risk	A	B	C	D	Significant Hazard
EXCAVATION WORK								
	Area clearing	Insects biting	Ill health	1	3	1	3	
	Mishandling of tools, core box during digging	Bruising by sharp edges	Cut / Crush injury	2	3	1	6	
	Fall of person	Fall hazard	Injury	1	2	1	2	
	Excavation work using JCB	Caught in between moving parts	Crush injury	3	1	2	6	
	Handling of breaker	Vibration	Ill health	1	2	1	2	
		Current Shock	Fatal	5	1	1	5	
CONSTRUCTION								
	Filling work	Mishandling of tools	Cut injury	1	4	1	4	
	Site levelling	Mishandling of tools	Crush injury	1	2	1	2	
	Footing concrete	Fall hazard	Injury	1	2	1	2	
	Column rising work	Fall hazard	Injury	2	2	1	4	
	Plinth beam concrete	Bruise of nail	Cut injury	1	3	1	3	
	Column rising work up to roof beam bottom	Fall hazard	Injury	5	1	3	15	Significant
	Roof concrete work	Fall of object, fall of person	Injury, fatal	1	2	1	2	

S. No	Activity	Hazard	Risk	A	B	C	D	Significant Hazard
	Brick work in ground floor	Dust generation	Ill health	1	3	1	3	
	Plastering work among the brick work	Splashing of cement mixture to eyes	Eye injury	1	2	1	2	
	Painting work	Fall hazard	Injury, fatal	5	1	2	10	
	Plumbing	Fall of objects / person, mishandling of tools	injury	5	1	2	10	
	Tiles laying	Slip hazard	Ill health	1	2	1	2	

D = A x B x C

Significant Hazard D ≥ 15

APPENDIX – 2 –

SAFE WORK PROCEDURE FOR WORKING AT HEIGHT > 1.8m

Date:

Permit no : WP- HW / SITE

Permit required for:

(Contractor details)

Location of work:

Nature of work:

Date of work:

Permit required: From-   To

Number of persons to work:

Signature   Signature

Name   Name

Contractor In charge for the Work: Supervisor / Engineer
In charge for the work

Permit to work issued by department:
Work started at:
Work completed at:
We have completed the work:
Contractor:

Engineer / Supervisor:
In charge of the work
Work permit closed & normality restored:

Permit Issuing department:
Working at height Hazards:
Fall from height, Slip hazard, Trip hazard, Hit against object, Electric shock

PPEs to use:
Safety Harness, Helmet, Safety Shoe, Tool Holder
Safety Precautions

1. Nobody shall be allowed to work at height (> 1.8 meter) without valid work permit and site supervisor to keep the permit while working.

2. Safety Harness shall be used by all workmen /supervisor, working at height.

3. Sufficient access ladders, work place scaffolding, work platforms shall be arranged properly.

4. Only trained manpower shall be used.

5. Persons working at >8 meter height are to be medically fit prior to start of work.

6. The Tools / Materials taken to work place at height shall be kept properly in a holder and shall be brought down after work is over with proper checking.

7. The concerned supervisor must be available at site during the period of work.

8. Electrical power supply, if it concerns the place of work, shall be isolated.

APPENDIX – 3 –

SAFE WORK PROCEDURE FOR ELECTRICAL WORK

Date of Issue: Valid from _____ hrs to _____ hrs

1. Detail of work and mains to be isolated:

Requested by:

Name _____   Time _____

Signature _____   Date _____

Designation

2. Equipment mentioned above has been effectively isolated from electrical source

by:-   Yes / No / NA

a) Removal of fuses

b) Putting of mains

c) Locking of mains & Tags displayed

d) Isolation tested and found OK

3. Any other Remarks:

Permit issued by: _____ Permit surrendered by: _____

APPENDIX – 4 –

SAFE WORK PROCEDURE FOR LIVE ELECTRICAL AREA

Date:

1. Location of Work :

2. Nature of Work :

3. Person and Department Requesting Work Permit:

4. Safety Precautions to be taken:

Approved by:

NAME & DEPARTMENT:

SIGNATURE:

DATE:

Person Receiving the Work Permit:

I have read and agree to follow the above special Safety precautions.

Name : _____ Signature: Date:

Surrender of Work Permit:

Name : _____ Signature: Date:

APPENDIX – 5 –

SAFE WORK PROCEDURE FOR HOT WORK

PERMIT - CUTTING, WELDING, GRINDING, ELECTRIC DRIVEN TOOLS & ENGINE DRIVEN TOOLS

Date of Issue: Valid from hrs to hrs

1. Permittee's Name: Agency / Department:

2. Exact description and location of work.

3. Precaution to be taken by the permittee.

Yes / No / NA

a) Floors kept clean of combustibles.

b) No combustibles or flammable liquids.

c) Combustible or flammable liquids covered with asbestos or metal shields.

d) All wall and floor openings covered.

e) Fire proof Tarpaulins suspended beneath work.

f) Grounding of welding machine

g) Fire watch available

h) Extinguishers available near to the welding work.

3. Remarks if Any:

4. Gas Test Values:

Oxygen level CO_2 Level Other Gas

5. Continuous inspection required: Yes No NA

Permit issued by

Site Safety Engineer Permit surrendered by _____

APPENDIX – 6

SAFE WORK PROCEDURE FOR EXCAVATION WORK & TRENCH Hazards

Fall, Trip hazard, Slip hazard, Hit by object, Hit against object, Electric shock, Insect bite

PPEs to use
Safety shoe/ Gumboots, Safety Helmet, Insulated Crow bar, Tool kit, Rubber hand gloves

Safety Precautions

1. No contractor shall be allowed excavation / digging work within construction premises without a valid work permit. Permit shall be available with the site supervisor at the time of execution of work.

2. Contractor shall instruct all workers to work strictly as per instruction & care shall be taken to ensure no damage to electrical cables / underground water/ gas pipe line etc.

3. Safety Barricade tape shall be placed around the trench at the time of work.

4. Proper care shall be taken to bring back all tools after the completion of work.

5. Electrical power supply shall be isolated as per the situation & use of Lock out /Tag out procedure to follow.

6. Civil Department to check the area fully before closing the permit.

7. The worker shall ensure proper shoring / planking of excavate, so that it will not create problem while working there.

APPENDIX – 7 –

SAFE WORK PROCEDURE FOR CONFINED SPACE ENTRY

Date of Issue: Valid from hrs to hrs

1. Permittee's Name: Agency / Department:

2. Exact description and location of work.

3. Preparation: Probability of special hazards from

a) Oxygen deficiency

b) Toxic gas/vapours/dust

c) Moving/rotating machinery

d) Electrical system

e) Hot liquids/vapours/chemicals

f) Dust heap

4. Precaution Taken Yes No NA

a) Oxygen testing carried out and found OK

b) Toxic gas/vapour analysis carried out and result is satisfactory

c) Electrical isolation done.

d) Observer poste d.

e) Vessel emptied out.

f) Effective isolation of pipe line done.

5. Precautions to be taken by the permittee:

a) The work to be carried out under supervision.

b) Use only 24 Volts lamps.

c) Portable electrical tools should have earthing.

d) Only trained personnel should do this work.

e) Others

6. Gas Test Values: Oxygen level

CO_2 Level

Other Gas

7. Continuous inspection required: Yes No NA

Permit issued by:

Permit surrendered by:

Descriptive Questions

1. Why Project specific Site Safety manual is required to be issued. Describe the various types of hazards in a project.
2. Describe the various reports generated by Safety department.
3. What is Safety Performance measurement? How it reflects the working conditions in a plant or a project?

Multiple Choice Questions

1. Lightening Arresters are required to be installed
   i. After completion of work
   ii. Before starting structural erection

State True or False

SL	Statement	True	False
01	You receive a consignment of new chain pulley block. You need to get it inspected by a Competent person before use		
02	You receive a new Hydra crane at site without registration number. It can be used immediately		
03	Every employee needs to undergo pre employment medical check up		
04	Female workers can work beyond 7 PM at site		
05	Near Miss reporting does not help an organisation		
06	A worker is injured during an accident at project and does not come to work for 3 days. No action need to be taken		
07	Waste water from Labour colony can be allowed to flow in adjacent farms without any problem		
08	Selection of Personnel Protection Equipment has no relation with activity being carried out		

SL	Statement	True	False
09	Water can be used to extinguish all types of fires		
10	Project site is deemed to be a Factory		
11	Noise from equipment needs to be monitored and controlled		
12	Risk assessment needs to be carried out for all activities		
13	Tool Box meeting is important in creating Safety awareness		
14	Every person entering the premises needs to be imparted Safety training		
15	Safety week is celebrated anytime in the year		

EMERGENCY RESPONSE PLAN

18.1 INDEX

1. INTRODUCTION

Industrial safety is a priority issue attracting everybody's concern in order to provide a Working environment, which is safe for the work force. A great deal of efforts and money is spent to reduce the scale & probability of hazards in the industries. However, there remains a finite possibility that certain hazards may occur. They can and have given rise to suffering and damage the extent of which is in part determined by the potential for loss, surrounding the event. Effective action has been possible in the emergency situation, due to existence of pre-planned and practiced procedure for dealing with such emergencies. The objective of the emergency plan is to define the action to be taken at departmental level and at project site level and these actions aim at the protection of the people and property within the works boundary and outside.

2. DEFINITION

A major emergency occurring at the works is one that may affect several departments within it and/or may cause serious injuries, loss of life, extensive damage to property or serious disruptions outside the works. The following essential things should be taken care-of for the effective implementation of the Emergency Preparedness Plan.

    i.   Periodic review and Updating of the plan.

    ii.   Periodic rehearsal of the plan by way of Mock Drills.

    iii.   Review and strengthening of the Resources needed.

    iv.   iv. Training of the site personnel in handling emergency equipment like use of various types of Fire-Fighting Equipment's.

    v.   Display of signage boards mentioning DO's & DON'Ts during any emergency at prominent places like

for Do's

    a.   Listen the emergency siren and assemble at the assembly point.

    b.   Obey the order of supervisor.

And For Don't

    a.   Violation of instruction of supervisor.

    b.   Continue to work after declaration of emergency.

3. BASIC OBJECTIVES OF EMERGENCY RESPONSE PLAN

i. Objective control and initial and Preventive control.

ii. Reduce risk to human health & life both within the works boundary and outside.

iii. Minimize damage to property both within and outside the boundary.

iv. Liaison effectively with the Government Authorities/Public/Press to avoid panic Situation and public disorder.

v. To protect the environment.

vi. To bring down the number of near-miss accidents to a minimum.

4. EMERGENCY

a. On Site Emergency- It is an emergency situation created by an incident that takes place in a plant or project site and effects are confined to the site premises involving the people working in that area.

b. Off Site Emergency –When an incident takes place in side of plant or project site and its effects are felt outside its premises, road accident, toppling of heavy vehicles etc., the situation thus created is called as "Off site emergency".

ESSENTIAL ELEMENTS OF THE EMERGENCY RESPONSE PLAN

i. Assignment of specific responsibilities to all categories of personnel (Main duties of personnel involved in emergencies are enumerated in Appendix-II).

ii. Effective communication of the nature of emergency to authorities within the organization.

iii. Institutions of works emergency action.

iv. Maintenance of orderly public relation.

v. Termination of Emergency.

vi. Appraisal of Local Hospitals/Nursing Homes/Doctors on specific emergency treatment required for persons affected, relevant to _____ site operations.

5. CRITERIA FOR PROCLAIMING MAJOR EMERGENCY

A major emergency situation can occur in the event of:-

    a.   Fire:-

        Fire can take place inside the Plant/Various departments/Stores/Paints, solvents and oil storage yards/offices/ and other areas inside the site.

    b.   Explosion:-

        Explosion followed by fire or fire followed by explosion. These could take place inside the Plant/Handling and filling of Hydrogen/ Storage Yards & Stores.

    c.   Release of Toxic/Corrosive Chemicals/ Radiations:-

        Though the development of major leaks and release of toxic/corrosive chemicals/ radiations create a major emergency situation in view of the limited quantities stored, the procedure outlined to handle the emergency situation is to be followed with respect to:-

    ·   Evacuation and assembly of personnel to a safe place.

    ·   Taking appropriate steps for bringing the situation under control.

6 ON SITE CENTERS FOR EMERGENCY CONTROL

The fundamental need of a Emergency Response Plan (ERP) is to establish two control centers, one at the affected site (Incident Control Centre) and the other one reasonably away from the affected site (Central Control Centre). The later should not be affected by the emergency situation. The Central Control Centre would be located at security room at main gate.

6.1 INCIDENT CONTROL CENTER (ICC)

This is a command post; this centre will seek the services of the emergency tasks force (Refer Appendix-III) and also contact the personnel listed in the Appendix-II for any assistance during the emergency. The important telephone numbers are listed in the Appendix-VI for contacting various essential services during the emergency. The residential address of important persons to be contacted for communicating the emergency situation is also listed in Appendix-VII. This centre is the communication link between the Emergency services and Central Control Centre. The Department-In-Charge takes control of this centre and acts as an Incidental Controller Location of ICC for different departments/areas and the respective incident controllers are given in Appendix-I. ICC Equipped with Telephones

	Name	Designation	Contact
1.			
2.			
3.			
4.			
5.			

6.2 CENTRAL CONTROL CENTre (CCC)

This centre is utilized for receiving and assessing information regarding the situation directing the resources to Incident Control Centre, at the demand of Incident Controller located at ICC, notifying neighbouring companies that might be affected, (Refer Appendix-V for neighbouring industries), calling in assistance from External Fire Services, ambulance etc. (Refer Appendix-IV for Outside Emergency Services). The Project site chief (Emergency Controller) is in charge of this centre, which will be located at security room at main gate and also liaisons with Department-In-Charge and Corporate office personnel according to need. CCC is Equipped with:-

    a.   Internal/External Telephone Nos.

    b.   Site-Master Plan

    c.   List of important persons in the site, head office, and authorities outside along with their names and telephone numbers.

7. HANDLING OF AN EMERGENCY SITUATION/DISASTER
7.1 RAISING AN ALARM:

In the event of emergency, the Department/Area-in-charge will inform the Security Office by phone (Phone No.– _____) or through messenger the guard on duty, who will raise an alarm, which should be different from fire).
DETAILS:-

Switch on the Siren (at the Security Office) and switch off after 5 seconds. Repeat 5 more times to make it 6 short duration sirens. The fire alarm would be raised from the fire control office located at Security office at main gate, as usual and would be a continuous siren. After raising alarm the Department/Area -In-Charge of the affected area will act as temporary Incident Controller of the situation. As soon as the Project in charge / Manager or his alternate arrives at the spot the latter takes the role of the Incident Controller at the Incident Control Center.

7.2 ASSESSMENT OF THE MAJOR EMERGENCY SITUATION:

Assessing the Major Emergency situation is to be done primarily by the concerned sectional Head /Manager. He assesses the situation and conveys information regarding the Emergency situation to the Project site chief/ Construction Manager and other important persons whose telephone numbers are given separately in Appendix-1. Project site chief/ Project Director in consultation with the Departmental Head of the affected site will take decision, on any additional resources required from outside in handling the situation, and assume the works of Emergency Controller (Project site chief) whose control point will be Central Control Centre located at security office at main gate. Depending on level of emergency his job will involve co-ordination of rescue and other emergency activities on site and liaison on with the respective authorities (Police, Fire Services, Hospitals & Press) for assistance in fire fighting, hospitalization etc., Fire Fighting Instructions & Assembly Point details are covered in Appendix-VIII & IX for the Emergency Controller to act upon and give instructions to the Task Force Member & other Emergency Services. When the conditions causing the Emergency are brought under control and the Project site's internal resources can tackle residual situation, Emergency

Controller will announce termination of Emergency situation.

The declaration of the end of the Emergency shall be made thus:-

    a.   Siren with continuous note will be sounded for 1 minute.

        Emergency Controller or his alternate shall communicate the termination of emergency situation to relevant authorities (Fire Services/Police/Local Authorities etc.)

    b.   If fire/ accident take place security personnel should be immediately cordoned the incident control centre as well as the main gate to avoid unwanted gathering/vehicles to maintain the law & order.

8. PREVENTIVE MEASURE TO AVOID ACCIDENTS & EMERGENCY SITUATIONS

Preventive measures for avoiding emergency situation in various work areas are prepared and issued to concerned Emergency sites (as shown in Appendix-XIV) the Do's & Don'ts slogans are affixed at appropriate work centers for adherence to avoid emergency situations.

8.1 HAZARDOUS AND TOXIC CHEMICALS

The hazardous and toxic chemicals used in the Owner are listed in the Appendix-X. The First-Aid measure to be followed at the time of incidents is covered in the information provided under each chemical along with other precaution to be followed while storing and handling such chemicals. The information provided under each chemical also includes the fire extinguishing media to be used along with steps to be taken at the time of spillage. The waste disposal method is also indicated against each hazardous chemicals covered under this category. The details are shown in Appendix-X and the MSDS book covering the details are issued to the concerned departments for immediate reference.

9. MOCK DRILL SCHEDULE FOR CHEMICAL/FIRE ACCIDENTS
9.1 CHEMICAL ACCIDENTS

Mock drills for chemical accidents shall be organized on quarterly basis in the emergency sites, (Emergency Sites listed in the Appendix-XIV) preferably during first half of the year. The mock drill information collected after the drill is evaluated for any corrective actions required. The training programs are organized where required.

9.2 FIRE FIGHTING DRILLS

In case of Fire fighting, on quarterly basis drills are organized to tackle the emergency situation. The fire extinguishers positioned in various locations for extinguishing different types of fire (Refer Code for Fire Safety issued by Head (Safety)) are shown in the Appendix-XIII.

10.   APPENDIXES :

S.No.	Title	Page No.
I	Emergency Controller and other officers for Assessing Major emergency Situation	08
II	Main Duties of Personnel Involved in Emergencies	09
III	Emergency Task Force	10
IV	List of outside emergency services	11
V	Important telephone numbers (Internal)	11
VI	Contact numbers of persons to be contacted in case of emergency	12
VII	Fire fighting instruction	12
VIII	Important points to be considered while carrying out fire fighting	13
IX	Assembly point	13
X	Guidelines to be followed to prevent/avoid accidents	13
XI	Fire extinguishers positioned in various departments/Areas	15
XII	Emergency sites identified in project site	16
XIII	Health hazards and safety risks in different stages of execution	17
XIV	On-Site safety requirements to be complied by all employees/workers/contractors	17
XV	Available emergency equipment	18
XVI	Site Plan	18
XVII	List of identified emergency situations	19
XVIII	DOs & DON'Ts during emergency	19
XIX	List of Contractors	19

APPENDIX –I

EMERGENCY CONTROLLER AND OTHER OFFICERS FOR ASSESSING MAJOR EMERGENCY SITUATION

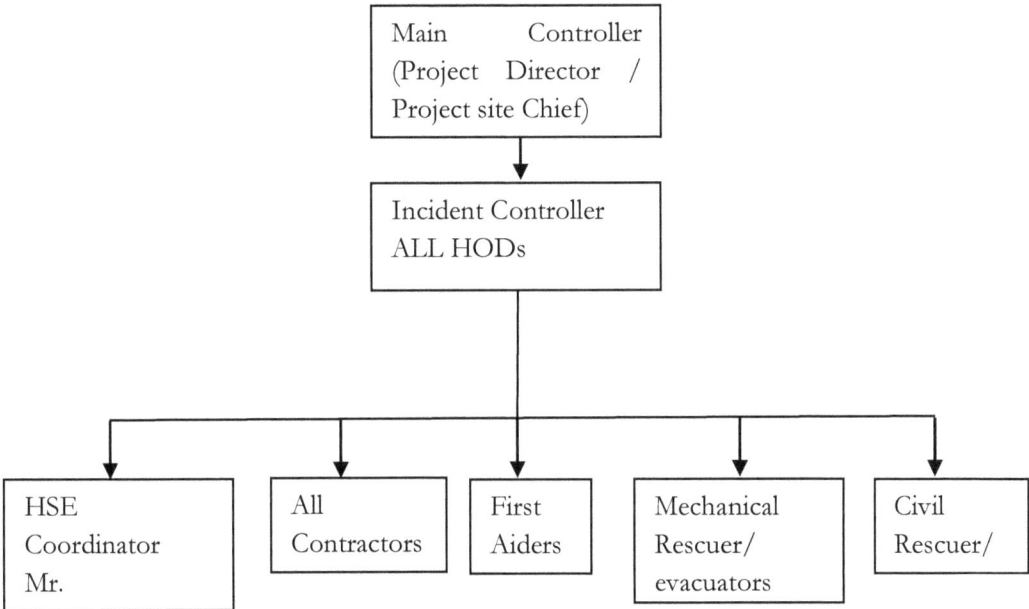

```
                    ┌─────────────────────────┐
                    │ Main        Controller  │
                    │ (Project  Director   /  │
                    │ Project site Chief)     │
                    └────────────┬────────────┘
                                 │
                                 ▼
                    ┌─────────────────────────┐
                    │ Incident Controller     │
                    │ ALL HODs                │
                    └────────────┬────────────┘
                                 │
        ┌────────┬───────────────┼───────────────┬────────────┐
        ▼        ▼               ▼               ▼            ▼
   ┌─────────┐ ┌──────────┐ ┌─────────┐ ┌────────────┐ ┌─────────┐
   │ HSE     │ │ All      │ │ First   │ │ Mechanical │ │ Civil   │
   │ Coord-  │ │ Contract-│ │ Aiders  │ │ Rescuer/   │ │ Rescuer/│
   │ inator  │ │ ors      │ │         │ │ evacuators │ │         │
   │ Mr.     │ │          │ │         │ │            │ │         │
   └─────────┘ └──────────┘ └─────────┘ └────────────┘ └─────────┘
```

ALL HODS:

1. HOD (CIVIL) –   Mr

2. HOD (MECHANICAL) –   Mr.

3. HOD (STORES / MATERIALS) – Mr.

4. HOD (ELECTRICAL) -   Mr.

5. HSE COORDINATOR –   Mr.

ALL CONTRACTORS:

List of contractors is mentioned in Appendix –

FIRST AIDERS:

Mr.

Mr.

MECHANICAL RESCUER/EVACUATORS:

Mr.

Mr.

Mr.

CIVIL RESCUER/ EVACUATORS:

Mr.

APPENDIX – II

MAIN DUTIES OF PERSONNEL INVOLVED IN EMERGENCIES

1. Person discovering Serious Unusual Occurrence/Fire
    I.   Raise fire alarm by pressing the call point in case of fire only. Fire alarm call points are located in Security Room. Major Emergency alarm could be raised only at the gate, under specific instructions from the Shift In-Charge and this alarm is different from fire alarm.

    II. Inform the Shift-In-Charge immediately about the incident.

    III. In the event of small fires fight the fire with the appropriate fire extinguisher.

    IV. Help the Shift-In-Charge as per his instructions.

2. Head of the Department
    I.   In the event of acting as Incident Controller carries out the function of Incident Controller. Quickly assess cause/source of the hazards and its effects.

    II. Discuss with the Project site chief / Safety Officer and coordinate the necessary action required to control/contain the emergency situation

III.   Continuously monitor the work of fire fighting personnel and other persons engaged in the emergency actions so that all actions are carried out safely.

3. Project Emergency Controller (Project site Chief)/ Construction Manager

I.   Relieve the Incident Controller of the responsibility of overall Main Control of Central Control Center (CCC).

II.   Discuss with the Incident Controller about the situation and evaluate the Major Emergency situation (The initial assessment of the situation is done by the Department-In-Charge immediately after the incident is found out). If the incident involved is a Major Fire, the Major Emergency situation is declared by the Department-In-Charge.

III.   Ensure communication to the following authorities:-

a. Fire Brigade if necessary

b. Local Hospitals

c. Civil Authorities

d. Electricity Board

Any one of the Emergency Task Force Personnel should contact the Fire Brigade.

I.   Ensure communication to the neighbouring industries of the incident.

II.   Maintain a speculative continuous review of possible developments and assess these to determine most probable course of events.

III.   Ensure proper preservation of evidence for subsequent investigation.

IV.   Inform the relevant company officers at the Regional/ Corporate Office.

4. HSE Coordinator/Safety Head

I.   Ensure overall Safety of the Emergency Operations.

II.   Assist in controlling the Emergency in connection with the plant personnel.

III.   Keep Central Control Centre (CCC) informed of the developments from time to time.

IV.   Mobilize all available resources for controlling the incident Help carrying out searches and rescue operations if required along with the plant personnel on receipt of information from incident Controller.

5. Contractors

    I.   Inform the Area-In-Charge immediately about the incident

    II.   Assist in controlling the Emergency in connection with the plant personnel

    III.   In the event of small fires fight the fire with the appropriate fire extinguisher

    IV.   Ensure communication to the neighbouring work areas of the incident

    V.   Train adequate persons in First aid and rescue operations

6. First Aiders

    I.   Assist rescuers and identify the type of first aid required

    II.   Organize first aid equipment

    III.   Give first aid as necessary

    IV.   Assist in sending victims to health centre speedily

---

APPENDIX – III

EMERGENCY TASK FORCE

The following members would constitute the Emergency Task Force of the site, and would compliment and supplement the efforts taken by Fire Brigade/Other Operations related to control of emergency

S. No.	NAME	INTERNAL TELEPONES	
		Office	Residence
1.			
2.			
3.			
4.			
5.			
6.			
7.			
8.			

APPENDIX – IV

LIST OF OUTSIDE EMERGENCY SERVICES -

A. FIRE STATION
- Fire Department – 101

B. POLICE STATIONS
- Local Police Station -

C. HOSPITALS AND AMBULANCE SERVICES
- Ambulance - 108
- Site Ambulance Driver-
- Local Hospital-

APPENDIX – V

IMPORTANT TELEPONE NUMBERS (INTERNAL)

IMPORTANT SERVICES INTERNAL

Security   Mr.

Fire   Mr.

First Aid   Male Nurse -

               Female Nurse –

ELECTRICAL

Power control room – Mr.

Electrical Complaints – Mr.

SECURITY

Main Gate No. _ – Mr.

APPENDIX – VI

CONTACT NO. OF PERSONS TO BE CONTACTED IN CASE OF AN EMERGENCY

1. PROJECT SITE CHIEF –   Mr.   No. _____

2. HOD (CIVIL) –   Mr.   No. _____

3. HOD (MECHANICAL) –   Mr.   No. _____

4. HOD (STORES / MATERIALS) – Mr.   No. _____

5. HSE COORDINATOR –   Mr.   No. _____

APPENDIX – VII

FIRE FIGHTING INSTRUCTIONS

The Site Fire-Fighting procedure gives details on fires/fire fighting instructions/location of various fire fighting equipment's etc. personnel are given practical training in fire fighting.

STEP BY STEP PROCEDURE TO BE FOLLOWED ON NOTICING A FIRE/ FIGHTING FIRES

1. Anyone noticing fire in any area of the works shall immediately inform the Security Office through the internal telephone No. ___ or by personally going to the Security Office, whichever is quicker.

2. In case of small fire select/use the suitable fire extinguishers and extinguish the fire (Everyone must get familiarized, on the various types of fire and the operating principles and instructions for each type of extinguisher).

3. In case of big fire where the use of the extinguisher is insufficient and where the water could be used, Hydrant points are to be used, and for this at least two persons are required. Fire Hydrants: Rush to the nearest hose station, use the key for opening the fire hose box and collect the hoses & nozzles. Connect the male coupling of the hose to the hydrant and spread the hose. Connect the nozzle to the other end of the hose. Hold the nozzle firmly, and ask the other man to open the hydrant valve

slowly. Direct the nozzle towards the base of the fire. In case one hose length is found insufficient another hose should be connected to the first hose and at the end of second hose the nozzle should be connected. The security guards on duty should rush at the fire place immediately and should help in fighting the fire.

4. For fighting fire involving oils and other hydrocarbons, ABC type fire extinguishers, which are installed at various sensitive locations clearly marked as FIRE, should be sprayed at the base of the fire & not water. If the fire is small then sand buckets can also be used

5. In case of fires contacting to the office of fire by contacting at phone Number 8349995683.

APPENDIX – VIII

IMPORTANT POINTS TO BE CONSIDERED WHILE CARRYING OUT FIRE FIGHTING/RESCUE OPERATIONS

1. DIRECTION OF SMOKE:-

a. Fight fire from the opposite direction of the wind flow.

2. CONSEQUENT CHEMICALS RELEASE:-

a. Take extra care like using gloves, safety shoes, masks goggles etc. to minimize effect.

3. RESCUING TRAPPED PERSONS:-

a. Enter smoke filled room with masks and a long rope tied around waist for rescuing trapped persons.

4. DEALING FIRE WITH WATER:

a. While fighting fire with water safeguard area in order to prevent contamination of Water sources.

APPENDIX - IX

ASSEMBLY POINT

There is only one assembly point in the site:
Assembly point near main gate (time office), always assemble at the main gate of the Project site in case of an emergency. Several assembly point boards posted at different locations.

APPENDIX - X

GUIDELINES TO BE FOLLOWED TO PREVENT/AVOID ACCIDENTS

A. SAFETY IN CHEMICAL STORE

1. Concentrated acids are to be handled with care as they can readily cause burn injuries. Direct contact with them should be avoided.

2. Scrupulous use of safety equipment's should be adopted.

3. Splash on bare skin should be immediately treated by flushing water on the point of control.

4. Whenever flammable chemicals are in use, proper ventilation is to be arranged. Exhaust fans should invariably be used.

5. Open flame, if any, around should be put off.

6. Use of proper firefighting equipment is to be known to every individual on the floor.

7. A very high standard of housekeeping is warranted in the chemical store. There should never be mix up.

8. Acid should be stored at ground level. Proper exhaust system and adequate illumination should be available.

9. A chemical without a label to be abandoned.

B. SAFETY IN BATTERY CHARGING

1. Store sulphuric acid in plastic container or glass container.

2. Always add acids slowly to the water when mixing, especially when diluting high gravity acid never add water to the acid.

3. Always wear acid & alkali proof goggles and rubber gloves when handling acids, be extremely careful not to spill or splash acid.

4. During charging of the battery connect the battery terminal correctly to the DC source. If the battery is connected incorrectly, then permanent damage may result.

5. The battery must always be recharged immediately after the complete discharge.

6. When adding distilled water, which has evaporated, never fill the cells above the normal level. Overfilling causes loss of acid thus reducing the battery capacity.

7. If battery acid gets in contact with your skin, flush it at once with plenty of water. Apply Baking Soda, if available, on the burnt area. Baking soda neutralizes the acid.

C. SAFETY IN ARC WELDING

Follow OCP

1. Electrodes should be removed from holders when not in use.

2. Always "Switch Off" the welding equipments after the use and disconnect the plug.

3. Never direct a "LASER" beam to a job, which reflects the beam back as reflection may result in losing eyesight.

4. Whenever Equipment is transported from place to place, take precautions.

5. Regularly inspect the insulation on electrode holder cables & accessories. Replace worn & damaged cables immediately.

6. Cover the lug terminals to prevent short-circuiting out by a metal object.

7. Ensure cable & power source are free from dirt/grease.

8. Remove combustible materials from work area or cover them with fire resistant blankets, before starting the welding operations.

9. Use all proposed safety/protective Equipments.

10. When job is finished, disconnect the welding machine from the power source and remove the electrode from the holder. Store the electrode in safe place.

D. SAFETY IN ELECTRICITY

1. Do not reuse a blown fuse until you are satisfied that the defect has been rectified.

2. Do not put-off any switch unless you are familiar with the circuit which it controls and know the reason for its being in "ON" position.

3. Do not touch or tamper with any electrical gear or conductor, unless you have made sure that it is dead and earthen.

4. Do not work on the live circuit without the express orders of the supervisors. Make certain that all safety precautions have been taken and you are accompanied by second person to render First-Aid and artificial respiration.

5. Do not connect earthling connection or render ineffective the safety gadgets installed on mains and apparatus.

6. Don't use or close switch or fuse slowly, do it quickly and positively.

7. Do not use wire/cables with poor insulation

8. Do not touch an electric circuit when your hands are wet or bleeding from a cut or an abrasion.

9. Do not work on energized circuits without taking extra precaution such as the use of rubber gloves.

10. Do not disconnect a plug by pulling a flexible cable when the switch is "ON".

11. Do not use fire extinguisher on electrical equipments unless it is switched "OFF".

12. Do not throw water on live electrical equipments in case of fire.

APPENDIX – XI

FIRE EXTINGUISHERS POSITIONED IN VARIOUS DEPARTMENT/AREAS

Sl. No.	Department/Area	CO_2	Foam	Sand Bucket 9 Ltrs.	ABC Type (5 Kgs.)	Total Extinguisher
1.					1	1
2.	Admin	1				1
3.					1	1
4.	Steel Yard				1	1
5.	Civil Lab	1				1
6.	Store				1	1
7.	Diesel Storage Area		4			4
8.	Control Room	2				2
9.	Fabrication Yard	2		1 set		3
10.	BOILER-	1		1 set		2
11.	Labour Colony	1				1
12.	C.W.S.T	1				1
13.	ESP	1				1
14.	Store Side				1	1
15.	(D.G) Office Side				1	1
16.	Chimney	1				1

17.	Staff Quarter	1				1
18.	Kitchen				1	1
19.	Staff Kitchen				1	1
20.	Sub Station			1 set		1
Total		12	4	3 sets	8	27

APPENDIX – XII

EMERGENCY SITES IDENTIFIED IN PROJECT SITE

Type of hazards that may occur

1. **BOILER** - FALL OF OBJECT OR MAN AT HEIGHT, SUPERHEATED STEAM LEAKAGE

2. **CONDENSER PIT** - FALL OF MAN IN THE PIT

3. **H2 FILLING STATION** - GAS LEAKAGE, EXPLOSION DURING THE

   FILLING

4. **TRANSFORMER YARD** - ELECTRICAL SHOCK, HIGH VOLTAGE

5. **FO TANK YARD** - FIRE

6. **WATER TREATMENT PLANT** - FALL OF MAN IN THE PIT

7. **COAL HANDLING PLANT** - FIRE

8. **BUNKERS** - FIRE

9. **ELECTRICAL TOOLS** - ELECTROCUTIONS AND FIRE.

10. **COOLING TOWER** – COLLAPS OF WALL

11. **CWST** – FALL OF MAN

12. **CHIMNEY** – COLLAPSE OF WALL

13. **ESP** – FALL OF MAN

14. **TG BUILDING** – COLLAPSE OF WALL

APPENDIX - XIII

HEALTH HAZARDS AND SAFETY RISKS IN DIFFERENT EXECUTION STAGES

These are as per the hazard identification and risk assessment carried out

Records / References:

1. Initial Environmental Review

2. Initial OHS review

3. Mock drill records

4. Safe work practices/ OCPs

APPENDIX - XIV

ON SITE SAFETY REQUIREMENTS TO BE COMPLIED BY ALL EMPLOYEES / WORKERS / CONTRACTORS

1. No person is allowed to work with state of alcohol / spirit consumption which may pose to his/her safety while at work.

2. No person is allowed to work, if he/she has consumed any prohibited drugs/narcotics.

3. No consumption of alcoholic drinks and prohibited drugs are allowed in the project premises.

4. All workers are required to report to respective In charges before starting in the work, in case he/she has been taken/administered any prescribed medicine, causing discomfort, drowsiness, lethargic physical condition, or any other kind of body/mental effects no conducive for safe work.

5. No employee shall enter to areas marked as "No Unauthorized entry" unless authorized by area in charge by entry in the register kept under his custody.

6. All employees shall follow the safety measures in particular area/machine as notified by Site in charge.

7. For working at height, climbing at elevated plat forms, performing maintenance and electrical repair, concerned employees need follow the safe work practices/ OCPs. Area in charges are required to strictly ensure this.

8. All employees to use necessary personnel protective equipment as documented in the operations control procedure or through contact with EHS coordinator.

9. No one to keep any obstruction on the walkway passage.

10. To notify HOS electrical any loose/damaged electrical wiring open switch boards & not to undertake any repair by himself / herself.

11. Concerned employee to keep the tag/ Display board on the M/C while it is taken for maintenance. The tag/ Display board to read as "Under Maintenance". No employee shall start/operate the m/c under such display as this may cause serious OHS Hazards.

12. Concerned employee must ensure the right placement of m/c guards before starting the operation. They must remove the m/c "Under Maintenance" card only after informing respective in charge.

13. All employees to necessarily report the occurrence of Incident/Accident in the prescribed form kept with EHS coordinator.

14. No unauthorized person shall walk/move in the inner side of Yellow border/ ropes drawn around the area.

15. If there is any storage area within/between m/c the same shall be marked with black strips.

16. No employee shall use any container/carboys without providing label on the same after informing/consulting Safety Officer/Manager

APPENDIX - XV

AVAILABLE EMERGENCY EQUIPMENTS

(List as applicable at site)

1. Alarm systems
2. Emergency lighting and power
3. Means of escape
4. Safe refuges
5. Critical isolation valves, switches and cut-outs
6. Fire fighting equipment
7. First aid equipment (including emergency showers, eye wash stations etc)
8. Communication facilities

APPENDIX - XVI

SITE PLAN (INDICATIVE)

Site plan is enclosed herewith.

APPENDIX – XVII

LIST OF IDENTIFIED EMERGENCY SITUATIONS

1. Acid spill during acid cleaning
2. Leakage of h2 / explosion during filling
3. Oil spill / fire during flushing
4. Oil spill/ fire during storage
5. Flooding of plant area
6. Caving in of excavated pits
7. Handling of chemicals
8. Superheated steam leakage
9. Fall from height
10. Fire
11. Structure collapse

APPENDIX – XVIII

DO'S & DON'TS DURING THE EMERGENCY

DOs	DON'Ts
1. Listen the emergency siren and assemble at the assembly point	1. Violation of instruction of supervisor
2. Obey the order of Supervisor	2. Continue to work after declaration of emergency

APPENDIX – XIX

LIST OF CONTRACTORS

S.No.	Name of Contractors
1	
2	
3	
4	
5	
6	
7	
8	
9	
10	
11	
12	

18.2   Descriptive Questions

1. Why Project specific Emergency Preparedness plan is required to be issued. Describe the various types of emergencies in a project.

18.3   Multiple Choice Questions

State True or False

SL	Statement	True	False
01	Water can be used to extinguish all types of fires		
02	Name of Persons to contact needs to be displayed prominently		
03	Mock drills are necessary to check preparedness		
04	Emergency preparedness plan is a confidential document		
05	Assembly points are mandatory		
06	Training on Emergency preparedness plan requirements needs to be imparted to all stakeholders		

CONSTRUCTION LAWS AND REGULATIONS

India has a federal structure of governance. The constitution has divided the subjects on which laws are passed between Central government and State governments. There are some subjects where both Central and State governments can enact a law; these are said to be on concurrent list.

There are three types of legislations

1. Acts

2. Rules

3. Codes – Indian Penal code (IPC), Civil procedure code (CPC) and Criminal procedure code (CrPC). Etc.

Generally Acts are passed by Legislature and outline what is desired. There are corresponding rules for the act that outline the procedure to achieve intent of Act. The rules specify the parameters to be achieved, structure of machinery to regulate the working and fees/ fines under the act.

In addition to the above codes, there are codes that are formulated by professional bodies and mandatory in working.

Salient features of the acts/ rules and codes applicable for construction work are listed for ready reference. Even though the title of act/ rule denotes a year of enactment, the latest amendments need to be taken into account. It should be noted that normally Central rules set the basic parameters, whereas individual states can fix the parameters more stringent than Central rules depending on the conditions in the state. It is important to have a library of applicable acts, rules and codes at site for ready reference.

Atomic Energy Act 1962 and rules

An Act to provide for the development, control and use of atomic energy for the welfare of the people of India and for other peaceful purposes and for matters connected therewith. Radiography source emits radiations and needs to be handled by authorised person only. Storing and transportation of radiographic source should be as specified by this act.

The Air (Prevention and Control of Pollution) Act and Rules

An Act to provide for the prevention, control and abatement of air pollution, for the establishment, with a view to carrying out the aforesaid purposes, of Boards, for conferring on and assigning to such Boards powers and functions relating thereto and for matters connected therewith.

Arbitration and Conciliation Act 1996

An Act to consolidate and amend the law relating to domestic arbitration, international commercial arbitration and enforcement of foreign arbitral awards as also to define the law relating to conciliation and for matters connected therewith or incidental thereto.

Battery Management and Handling Rules, 2001

These rules shall apply to every manufacturer, importer, re-conditioner, assembler, dealer, recycler, auctioneer, consumer and bulk consumer involved in manufacture, processing, sale, purchase and use of batteries or components thereof.

Responsibilities of consumer or bulk consumer.-
(1) It shall be the responsibility of the consumer to ensure that used batteries are not disposed of in any manner other than depositing with the dealer, manufacturer, importer, assembler, registered recycler, and reconditioner or at the designated collection centres.

(2) It shall be the responsibility of the bulk consumer to -

(i) Ensure that used batteries are not disposed of in any manner other than by depositing with the dealer/manufacturer/registered recycler/ importer/ reconditioner or at the designated collection centres,- and

(ii) File half-yearly return in Form VIII to the State Board.

(3) Bulk consumers to their user units may auction used batteries to registered recyclers only.

The Bonded Labour system (Abolition) Act, 1976

An Act to provide for the abolition of bonded labour system with a view to preventing the economic and physical exploitation of the weaker sections of the people and for matters connected therewith or incidental thereto.

Building and Other Construction worker's Act 1996

An Act to regulate the employment and condition of service of buildings and other construction workers and to provide for their safety, health and welfare measures and for other matters connected therewith or incidental thereto.

Central Sales Tax Act 1956 –

➢ Taxes on sales or purchases of goods in the course of inter-state trade or commerce were brought expressly within the purview of the legislative jurisdiction of Parliament.

➢ Restrictions could be imposed on the powers of State legislatures with respect to the levy of taxes on the sale or purchase of goods within the State where the goods are of special importance in inter-state trade or commerce.

The Child Labour (Prohibition and Regulation) Act, 1986

An Act to prohibit the engagement of children in certain employments and to regulate the conditions of work of children in certain other employments.

Companies Act 1956 and 2013

The Act in a comprehensive form purports to deal with relevant themes such as investor protection, inclusive agenda, fraud mitigation, internal control, director responsibility and efficient restructuring. The Act is also quite outward looking and in several areas attempts to harmonize with international requirements.

Contract Labor (Regulation and Abolition) Act 1970 and Rules

An Act to regulate the employment of contract labour in certain establishments and to provide for its abolition in certain circumstances and for matters connected therewith.

Employees Provident Fund Act and Rules

An Act to provide for the institution of provident funds, pension fund and deposit-linked insurance fund for employees in factories and other establishments.

Employees State Insurance Act 1948

to provide for certain benefits to employees in case of sickness, maternity and employment injury and to make provision for certain other matters in relation thereto.

Environmental (Protection) Rules-"Environmental Standards"

The Central Pollution Control Board (CPCB) has developed National Standards for Effluents and Emission under the statutory powers of the Water (Prevention and Control of Pollution) Act, 1974 and the Air (Prevention and Control of Pollution) Act, 1981. These standards have been approved and notified by the Government of India, Ministry of Environment & Forests, under Section 25 of the Environmental (Protection) Act, 1986.

Equal Remuneration Act, 1976

Payment of remuneration at equal rates to men and women workers and other matters

The Explosive Act and Rules

to regulate the manufacture, possession, use, sale, transport and importation of explosives.

The Explosive Substance Act

The expression "explosive substance" shall be deemed to include any materials for making any explosive substance; also any apparatus, machine, implement or material used, or intended to be used, or adapted for causing, or aiding in causing, any explosion in or with any explosive substance; also any part of any such apparatus, machine or implement.

Factories Act 1948

An Act to consolidate and amend the law regulating labour in factories.

Gas Cylinder Rules

Filling, possession, import and transport of cylinders: - (1) No person shall fill any cylinder with any compressed gas or import, possess or transport any cylinder so filled or intended to be filled with such gas unless:-

(a) such cylinder and its valve have been constructed to a type and standard specified in Schedule I as amended from time to time by an order issued by the Chief Controller;

(b) the test and inspection certificates issued by the inspecting authority in respect of cylinder and its valve are made available to the Chief Controller and prior approval of the said authority is obtained.

The Hazardous Wastes (Management and Handling) Rules -

Indian Contract Act 1872 –

It determines the circumstances in which promises made by the parties to a contract shall be legally binding on them. All of us enter into a number of contracts everyday knowingly or unknowingly. Each contract creates some rights and duties on the contracting parties. Hence this legislation.

Income Tax Act

An Act to consolidate and amend the law relating to income-tax and super-tax

Central Excise Act 1944

An Act to consolidate and amend the law relating central duties of excise on goods manufactured or produced in India.

Customs Act 1962

An Act to consolidate and amend the law relating to Custom duty.

Indian Boiler Regulation Act and Rules

Indian Boiler Regulations are the standards in respect of materials, design and construction, inspection and testing of boilers and boiler components for compliance by the manufacturer's and users of boilers in the country. These regulations are being updated regularly by amending them in line with fast changes in boiler technology by the Central Boilers Board.

Indian Electricity Act 2013 and Rules –

The act covers major issues involving generation, distribution, transmission and trading in power.

Indian Stamp Act 1899

a fiscal statute laying down the law relating to tax levied in the form of stamps on instruments recording transactions & Stamp duties on instruments specified in Entry 91 of the Union List (viz. Bills of Exchange, cheques, promissory notes, bills of lading, letters of credit, policies of insurance, transfer of shares, debentures, proxies and receipts) is levied by the Union. Stamp duties on instruments other than those mentioned in Entry 91 of the Union List above are levied by the States as per Entry 63 of the State List.

The Industrial Disputes Act 1948

An Act to make provision for the investigation and settlement of industrial disputes, and for certain other purposes.

Industrial Employment (Standing Orders) Act

An Act requiring employers in industrial establishments formally to define conditions of employment under them. It is expedient to require employers in industrial establishments to define with sufficient precision the conditions of employment under them and to make the said conditions known to workmen employed by them.

Inter-state Migrant Workmen (Regulation of Employment and Conditions of Service) Act 1979

This Act protects the dignity of life and the interests of the workers whose services are requisitioned outside their native states in India.

Limitation Act 1963

An Act to consolidate and amend the law for the limitation of suits and other proceedings and for purposes connected therewith.

Minimum Wages Act

An Act to provide for fixing minimum rates of wages in certain employments. These rates differ in states and depend on the location of site. The rates are published in gazette.

Mines and minerals (development and regulation) act, 1957

Motor Vehicle Act & Rules

regulates all aspects of road transport vehicles. The Act provides in detail the legislative provisions regarding licensing of drivers/conductors, registration of motor vehicles, control of motor vehicles through permits, special provisions relating to state transport undertakings, traffic regulation, insurance, liability, offences and penalties, etc.

National Building Code 2005

issued by BIS is a comprehensive building Code, a national instrument providing guidelines for regulating the building construction activities across the country. The comprehensive NBC 2005 contains 11 Parts some of which are further divided into Sections totalling 26 chapters. The salient features of the NBC include the changes especially in regard to further enhancing our response to meet the challenges posed by natural calamities and reflecting the state-of-the-art and contemporary applicable international practices.

The Noise Pollution (Regulation & Control) Rules, 2000 -

Payment of Bonus Act

An Act to provide for the payment of bonus to persons employed in certain establishments on the basis of profits or on the basis of production or productivity and for matters connected therewith.

The Payment of Gratuity Act 1972

The Act provides for the payment of gratuity to workers employed in every factory, shop & establishment or educational institution employing 10 or more persons on any day of the preceding 12 months. All the employees irrespective of status or salary are entitled to the payment of gratuity on completion of 5 years of service. In case of death or disablement there is no minimum eligibility period. The amount of gratuity payable shall be at the rate of 17 days wages based on the rate of wages last drawn, for every completed year of service. **Formula is - Last Wages *15*No. of years of services/26**

Payment of Wages Act 1964

The Payment of Wages Act was enacted with a view to ensuring that wages payable to employed persons covered by the Act were disbursed by the employers within the prescribed time limit and that no deductions other than those authorised by law were made by them.

The Petroleum Act and Rules

An Act to consolidate and amend the law relating to the import, transport, storage, production, refining and blending of petroleum.

The Private Security Agencies (Regulation) Act, 2005

An Act to provide for the regulation of private security agencies and for matters connected therewith or incidental thereto.

The Professional Tax and Rules

Professional Tax is the tax charged by the state governments in India. Any one earning an income from salary or any one practising a profession such as chartered accountant, lawyer, doctor etc. are required to pay this professional tax. The maximum amount payable per year is Rs.2,500/-(and in line with salary.)

The Public Liability Insurance Act and Rules 1991

An Act to provide for public liability- insurance for the purpose of providing immediate relief to the persons affected by accident occurring while handling any hazardous substance and for matters connected therewith or incidental thereto.

Radiation Protection Rules 2004

These rules shall apply to practices adopted and interventions applied with respect to radiation sources.

Service Tax Act 1994

principles for determining the modalities of levying the Service Tax by the Central Govt. and collection of the proceeds thereof by the Central Govt. and the State.

Shops and Establishment Act

The object of this Act is to provide statutory obligation and rights to employees and employers in the unauthorized sector of employment, i.e., shops and establishments. This Act is applicable to all persons employed in an establishment with or without wages, except the members of the employer' family.

The Standards of Weights and Measures Act and Rules 1976

Establishment of the weights and measure based on the SI units. Provides to prescribe specification of measuring instruments used in commercial transaction, industrial production and measurement involved in public Health and Human safety. The specifications are given in **the Standard of weights and Measures (General) Rules 1987**.

The Water (Prevention and Control of Pollution) Act and Rules

An Act to provide for the prevention and control of water pollution and the maintaining or restoring of wholesomeness of water, for the establishment, with a view to carrying out the purposes aforesaid, of Boards for the prevention and control of water pollution, for conferring on and assigning to such Boards Powers and functions relating thereto and for matters connected therewith.

The Water (Prevention and Control of Pollution) Cess Act

An Act to provide for the levy and collection of a cess on water consumed by persons carrying on certain industries and by local authorities, with a view to augment the resources of the Central Board and the State Boards for the prevention and control of water pollution constituted under the **Water (Prevention and Control of Pollution) Act, 1974**.

Workmen Compensation Act - The Workmen's Compensation Act 1923

was enacted to help workmen face the hardships resulting from accidents. These legal provisions apply equally to women workers also. An employer is liable to provide monetary compensation to a disabled workman, or to his dependents, in case of his death, if the disablement or death occurs "out of and in the course of employment."

Works Contract Tax Act 1994

Service tax on works contract was imposed on 1-6-2007. However, it was restricted works contract relating to immovable property. Its scope has been extended to works contract of movable property also w.e.f. 1-7-2012.

19.1   Descriptive Questions

1.   Describe the various types of statutes. How these affect progress in a project.

19.2   Multiple Choice Questions

1.   As per Indian constitution Acts for subjects in concurrent list are passed by
     i.   Parliament only
     ii.   State Legislature only
     iii. Both

2.   Used Batteries can be disposed off
     i.   In any way
     ii.   To a scrap dealer
     iii. To a Battery manufacturer or Battery dealer

3.   Maintenance of Electrical installation can be performed by
     i.   Anybody
     ii.   Engineer
     iii. Electrician
     iv.   Licensed electrician

4.   Provident Fund needs to be recovered from wages of
     i.   Every Employee

      ii.   Workers

     iii.   Supervisors

     iv.   Executives

     v.   Employees earning less than specified amount

5. For cutting trees on the construction site
   i. No permission is required
   ii. Permission from authority is required

6. Drinking water needs to be provided in construction site
   i. At one point
   ii. At every place where work is in progress
   iii. Not necessary, workers can bring their requirement

State true or false

SL	Statement	True	False
01	Atomic Energy Act is not applicable to Thermal Projects		
02	You receive a consignment of new chain pulley block. You need to get it inspected by a Competent person before use		
03	You receive a new Hydra crane at site without registration number. It can be used immediately		
04	Every employee needs to undergo pre employment medical check up		
05	Female workers can work beyond 7 PM at site		
06	Creche needs to be provided if female workers are employed		
07	Anybody can carry a radiography source from one place to another		

SL	Statement	True	False
08	A worker is injured during an accident at project and does not come to work for 3 days. No action need to be taken		
09	Waste water from Labour colony can be allowed to flow in adjacent farms without any problem		
10	Selection of Personnel Protection Equipment has no relation with activity being carried out		
11	We can transport as much diesel as we wish		
12	You can not engage migrant workers at project site		
13	Project site is deemed to be a Factory		
14	Service Tax is applicable on Manufactured items		
15	Noise from equipment needs to be monitored and controlled		
16	Royalty needs to be paid for civil works inputs		
17	Work contract tax is charged on Mechanical erection		
18	Any agency can be engaged for providing security services		

THE BUILDING AND OTHER CONSTRUCTION WORKERS

(REGULATION OF EMPLOYMENT AND CONDITIONS OF SERVICE) ACT, 1996

CHAPTER VI

HOURS OF WORK, WELFARE MEASURES AND OTHER CONDITIONS OF SERVICE OF BUILDING WORKERS

---

28. Fixing hours for normal working day, etc.-

(1) The appropriate Government may, by rules.-

(a) fix the number of hours of work which shall constitute normal working day for a building worker, inclusive of one or more specified intervals;

(b) Provide for a day of rest in every period of seven days which shall be allowed to all building workers and for the payment of remuneration in respect of such days of rest;

(c) Provide for payment of work on a day of rest at a rate not less than the overtime rate specified in section 29.

(2) The provisions of sub-section (1) shall, in relation to the following classes of building workers, apply only to such extent, and subject to such conditions, as may be prescribed, namely:-

(a) Persons engaged on urgent work, or in any emergency which could not have been foreseen or prevented;

(b) Persons engaged in a work in the nature of preparatory or complementary work which must necessarily be carried on outside the normal hours of work laid down in the rules;

(c) Persons engaged in any work which for technical reasons has to be completed before the day is over:

(d) Persons engaged in a work which could not be carried on except at times dependant on the irregular action of natural forces.

COMMENTS

Government has been empowered to fix the number of hours of work for a building worker, to provide for day of rest in every period of 7 days and for the payment of remuneration in respect of such days of rest, to provide for payment of work on a day of rest at a rate not less than the overtime rate.

---

29. Wages for overtime work.-

(1) Where any building worker is required to work on any day in excess of the number of hours constituting a normal working day he shall be entitled to wages at the rate of twice his ordinary rate of wages.

(2) For the purposes of this section, "ordinary rates of wages" means tile basic wages plus such allowances as tile worker is for the time being entitled to but does not include any bonus.

COMMENTS

If any building worker is required to work on any day in excess of tile number of hours constituting a normal working day, he is entitled to wages at the rate of twice his ordinary rate of wages.

---

30. Maintenance of registers and records.-

---

(1) Every employer shall maintain such registers and records giving such particulars of building workers employed by him, the work performed by them, the number of hours of work which shall constitute a normal working day for them, in day of rest in every period of seven days which shall be allowed to them, tile wages paid to them, the receipts given by them and such other particulars in such form as my be prescribed.

(2) Every employer shall keep exhibited, in such manner as may be prescribed, in tile place where such workers may be employed, notices in the prescribed form containing the prescribed particulars.

(3) The appropriate Government may, by rules, provide for tile issue of wage books or wage slips to building workers employed in an establishment and prescribe tile manner in which entries shall be made and authenticated in such wage books or wage slips by the employer or his agent.

---

31. Prohibition of employment of certain persons in certain building or other construction work.-

---

No person about whom the employer knows or has reason to believe that he is a deaf or he has a defective vision or he has a tendency to giddiness shall be required or allowed to work in any such operation of building or other construction work which is likely to involve a risk of any accident either to the building worker himself or to any other person.

COMMENTS

Any person who is deaf or who has detective vision or who has a tendency to giddiness is not required or allowed to work in any such operation of building or other construction work which is likely to involve risk of an accidents.

---

32. Drinking water.-

---

(1) The employer shall make in every place where building or other construction work is in progress, effective arrangements to provide and maintain at suitable points conveniently situated for all persons employed there in, a sufficient supply of wholesome drinking water.

(2) All Such points shall be legible marked "Drinking Water" in a language understood by a majority of the person employed in such place and no such point shall be situated within six metres of any washing place, Urinal or latrine.

33. Latrines and urinals.-

In every place where building or other construction work is carried on, the employer shall provide sufficient latrine and urinal accommodation of such types as may be prescribed and they shall be so conveniently situated as may be accessible to the building workers at all times while they are in such place:

Provided that it shall not be necessary to provide separate urinals in any place where less than fifty persons are employed or where the latrines are connected to a water-borne sewage system.

34. Accommodation.-

(1) The employer shall provide, free of charges and within the work site or as near to it as may be possible temporary living accommodation to all building workers employed by him for such period as the building or other construction work is in progress.

(2) The temporary accommodation provided under sub-section (1) shall have separate cooking place bathing, washing and lavatory facilities

(3) As soon as may be, after the building or other construction work is over, the employer shall, at his own cost, cause removal or demolition of the temporary structures erected by him for the purpose of providing living accommodation, cooking place or other facilities to the building workers as required under sub-section (1), and restore the ground in good level and clean condition.

(4) In case an employer is given. any land by a Municipal Board or by other local authority for the purposes of providing temporary accommodation for the building workers under this section, he shall as soon as, may be after the construction work is over, return the possession of such land in the same condition in which he received the same.

35. Creches.-

(1) In every place where in more them fifty female building workers are ordinarily employed, there shall be provided and maintained, a suitable room or rooms for the use of children under the age of six years, of such female workers.

(2) Such rooms shall-

(a) Provide adequate accommodation:

(b) Be adequately lighted and ventilated;

(c) Be maintained in a clean and sanitary condition;

(d) Be under the charge of women trained in the care of children and infants.

36. First-aid.-

Every employer shall provide in all the places where building or other construction work is carried on such first-aid facilities as may be prescribed.

37. Canteens, etc.-

The appropriate Government may, by rules require the employer-

(a) To provide and maintain in every place wherein not less than two hundred and fifty building workers are ordinarily employed, a canteen for the use of the workers;

(b) To provide such other welfare measures for the benefit of building workers as may be prescribed.

CHAPTER VII

SAFETY AND HEALTH MEASURES

38. Safety Committee and safety officers.-

(1) In every establishment wherein five hundred or more building workers are ordinarily employed, the employer shall constitute a Safety Committee consisting of such number of representatives of the employer and the building workers as may be prescribed by the State Government:

Provided that the number of persons representing the workers, shall, in no case, be less than the persons representing the employer.

(2) In every establishment referred to in sub-section (1), the employer shall also appoint a safety officer who shall possess such qualifications and perform such duties as may be prescribed.

39. Notice of certain accidents.-

(1) Where in any establishment an accident occurs which causes death or which causes any bodily injury by reason of which the person injured is prevented from working for a period of forty-eight hours or more immediately following the accident, or which is of such a nature as may be prescribed, the employer shall give notice thereof to such authority, in such form and within such time as may be prescribed.

(2) On receipt of a notice under sub-section (1) the authority referred to in that sub-section may make such investigation or inquiry as it considers necessary.

(3) Where a notice given under sub-section (1) relates to an accident causing death of five or more persons, the authority shall make an inquiry into such accident within one month of the receipt of the notice.

40. Power of appropriate Government to make rules for the safety and health of building workers.-

(1) The appropriate Government may, by notification, make rules regarding the measures to be taken for the safety and health of building workers in the

course of their employment and the equipment and appliances necessary to be provided to them for ensuring their safety, health and protection, during such employment.

(2) In particular, and without prejudice to the generality of the foregoing power, such rules may provide for all or any of the following matters, namely:-

(a) the safe means of access to, and the safety of, any working place, including the provision of suitable and sufficient scaffolding at various stages when work cannot be safely done from the ground or from any part of a building or from a ladder or such other means of support;

(b) the precautions to be taken in connection with the demolition of the whole or any substantial part of a building or other structure under the supervision of a competent person child the avoidance of danger from collapse of any building or other structure while removing any part of the framed building or other structure by shoring or otherwise;

(c) the handling or use of explosive under the control of competent persons so that there is no exposure to the risk of injury from explosion or from flying material;

(d) the erection installation, use and maintenance of transporting equipment, such as locomotives, trucks, wagons and other vehicles and trailers and appointment of competent persons to drive or operate such equipment;

(e) the erection, installation, use and maintenance of hoists, lifting appliances and lifting gear including periodical testing and examination and heat treatment where necessary, precautions to be taken while raising or lowering loads, restrictions on carriage of persons and appointment of competent persons on hoists or other lifting appliances;

(f) the adequate and suitable lighting of every workplace and approach thereto, of every place where raising or lowering operations with the use of hoists, lifting appliances or lifting gears are in progress and of all openings dangerous to building workers employed;

(g) The precautions to be taken to prevent inhalation of dust, fumes, gases or vapours during any grinding, cleaning, spraying or manipulation of only material and steps to be taken to secure and maintain adequate ventilation of every working place or confined Space;

(h) The measures to be taken during stacking or unstacking, stowing or unstowing of materials or goods or handling in connection therewith;

(i) the safeguarding of machinery including the fencing of every fly-wheel and every moving part of prime mover and every part of transmission or other machinery, unless it is in such a position or of such construction as to be safe to every worker working only of the operations and as if it were securely fenced:

(j) the safe handling and use of plant, including tools and equipment operated by compressed air:

(k) the precaution to be taken in case of fire;

(l) the limits of weight to be lifted or moved by workers;

(m) the safe transport of workers to or from any workplace by water and provision of means for rescue from drowning;

(n) the steps to be taken to prevent danger to workers from live electric wires or apparatus including electrical machinery and tools and from overhead wires;

(o) the keeping of safety nets, safety sheets and safety belts where the special nature or the circumstances of work render them necessary for the safety of the workers;

(p) The standards to be complied with regard to scaffolding, ladders and stairs, lifting appliances. Ropes, chains and accessories, earth moving equipment and floating operational equipment's;

(q) the precautions to be taken with regard to pile driving, concrete work, work with hot asphalt, tar or other similar things, insulation work, demolition operations, excavation, underground construction and handling materials;

(r) the safety policy, that is to say, a policy relating to steps to be taken to ensure the safety and health of the building workers, the administrative arrangements (there for) and the matters connected therewith, to be framed by the employers and contractors for tile operations to be carried on in a building or other construction work:

(s) the information to be furnished to the Bureau of Indian Standard established under the Bureau of Indian Standards Act, 1986 (63 of 1986), regarding the use of any article or process covered under that Act in a building or other construction work:

(t)   the provision and maintenance of medical facilities for building workers;

(u)   any other matter concerning the safety and health of workers working in any of the operations being carried on in a building or other construction work.

**MINISTRY OF LABOUR
AND SOCIAL SECURITY**

F1/10

```
┌─────────────────────────────────────────────┐
│                                               │
│            REPORT OF ACCIDENT                 │
│                 AS PER                        │
│        THE FACTORIES ACT. VOL V1              │
│                AS AMENDED                      │
│                                               │
└─────────────────────────────────────────────┘
```

SECTION 21 OF ACT

PART A

To be forwarded immediately (see page 3)

1. Name and Address of Factory or Works………………………………………..

2. Processes or
   Products……………………………………………………………………...

3. Date of accident……………………………….Time…………………..Day of
   Week…………………

4. Department which accident occurred………………………………………

5. Exact location in the Department where the accident
   occurred……………………………………………………………………...

6. Give the following details in respect of the injured person:-

(a) Name…………………………………………………………………………

(b) Home Address………………………………………………………………...

(c) Sex……………………………………………..Age……………………

(d) Normal Occupation...

(e) Occupation at time of accident if different from (d)
above...

(f) Time at which work was commenced on day of
accident..

7. How did the accident happen?...

....

8. What was the injured person doing at the time of the
accident?..

9. If caused by machinery:

(a) Give name of the machine and part causing injury...............................

....

(b) Was the machine being moved by mechanical power at the time of the
accident?......................................

(c) Was part causing injury guarded at the time of
accident?...

10. What injuries or damage resulted from the accident (e.g. fatal, loss of finger,
fracture of leg, scald, scratch followed by sepsis
etc.)...

11. Was first-aid treatment rendered and by
whom?..

12. Has the accident been investigated by Management?.....................

13. Name and substantive post of Safety Supervisor appointed under Reg. 57

....

....

14. Any other comments..

Signature.......................................

Manager or Person in Control of Factory or Works

Date.......................................

REPORT OF ACCIDENT

PART B

To be forwarded to the Chief Factory Inspector when worker has resumed.

1. Name and Address of Factory or Works.....................................

....

2. Name of Injured person...

3. Date of Accident...Time................

4. Date of resumption of work..

5. No. of days during which injured person was prevented by the injury from earning full wages at his normal occupation ....

....

Signature.............................

Manager or Person in Control of Factory or Works

Date.................................

THE FACTORIES ACT, VOL. VI

THE FACTORIES ACT

SECTION 21

"(1) Where any accident occurs in a factory which either -

(a) cases loss of life to a person employed in the factory; or

(b) disables any such person for more than two days from earning full wages at the work at which he was employed, the Manager of the factory or person having control of the machinery in such factory shall forthwith report the occurrence of such accident to the Chief Factory Inspector and in connection therewith he shall furnish such particulars as the Chief Factory Inspector in any case from time to time require.

(2) The Manager of the factory or person having control of the machinery as aforesaid shall also from time to time in like manner report to the Chief Factory Inspector –

(a) all accidental fires and explosions;

(b) the collapse or failure of any building or structure;

(c) accidents to machinery or plant which result in the cessation of work beyond the shift or day on which the accident occurs;

(d) any industrial disease which may be prescribed by the Chief Factory Inspector, which may occur in the factory."

NOTE:- The provisions of subsections (1) and (2) quoted above also apply to:

(a) building operations undertaken by way of trade or business, or for the purpose of any industrial or commercial undertaking, and any line or siding which is used in connection therewith and for the purposes thereof.

G.P.O

Notes

Notes

www.ingramcontent.com/pod-product-compliance
Lightning Source LLC
Chambersburg PA
CBHW061741210326
41599CB00034B/6749